Springer Theses

Recognizing Outstanding Ph.D. Research

Aims and Scope

The series "Springer Theses" brings together a selection of the very best Ph.D. theses from around the world and across the physical sciences. Nominated and endorsed by two recognized specialists, each published volume has been selected for its scientific excellence and the high impact of its contents for the pertinent field of research. For greater accessibility to non-specialists, the published versions include an extended introduction, as well as a foreword by the student's supervisor explaining the special relevance of the work for the field. As a whole, the series will provide a valuable resource both for newcomers to the research fields described, and for other scientists seeking detailed background information on special questions. Finally, it provides an accredited documentation of the valuable contributions made by today's younger generation of scientists.

Theses are accepted into the series by invited nomination only and must fulfill all of the following criteria

- They must be written in good English.
- The topic should fall within the confines of Chemistry, Physics, Earth Sciences, Engineering and related interdisciplinary fields such as Materials, Nanoscience, Chemical Engineering, Complex Systems and Biophysics.
- The work reported in the thesis must represent a significant scientific advance.
- If the thesis includes previously published material, permission to reproduce this must be gained from the respective copyright holder.
- They must have been examined and passed during the 12 months prior to nomination.
- Each thesis should include a foreword by the supervisor outlining the significance of its content.
- The theses should have a clearly defined structure including an introduction accessible to scientists not expert in that particular field.

More information about this series at http://www.springer.com/series/8790

Joshua D. Vande Hey

A Novel Lidar Ceilometer

Design, Implementation and Characterisation

Doctoral Thesis accepted by
Loughborough University, UK

Springer

Author
Dr. Joshua D. Vande Hey
Department of Physics and Astronomy
University of Leicester
Leicester
UK

Supervisor
Prof. Jeremy Coupland
Wolfson School of Mechanical and Manufacturing Engineering
Loughborough University
Loughborough
UK

ISSN 2190-5053 ISSN 2190-5061 (electronic)
ISBN 978-3-319-12612-8 ISBN 978-3-319-12613-5 (eBook)
DOI 10.1007/978-3-319-12613-5

Library of Congress Control Number: 2014953307

Springer Cham Heidelberg New York Dordrecht London

Printed on acid-free paper

Springer is part of Springer Science+Business Media (www.springer.com)

Supervisor's Foreword

It is with a sense of pride and satisfaction that I write this foreword to Josh Vande Hey's thesis. Like all good Ph.D. students, Josh has taught me a great deal and I believe his work serves as an excellent example of the engineering design process. It is applied science, it is driven by the needs of industry, it is inventive and it has resulted in a successful commercial product, the CS135 Lidar Ceilometer.

The project began in 2007—a collaboration between Campbell Scientific Ltd. and Loughborough University, funded by the Knowledge Transfer Partnership of the Technology Strategy Board. The aim of this project was to develop a new design of ceilometer—an instrument used by meteorologists to measure the height of cloud layers. In essence, a ceilometer is a lidar system that retrieves this information from the backscattered light that is collected when a laser pulse train is made to propagate through the atmosphere.

Like many other design projects, it began with a design brief that could be summed up as "build an instrument that does more and costs less than the competition". In this case, "more" meant the ability to reveal more detailed images of thin cloud layers, aerosols and airborne pollutants that are present in the lower atmosphere. At the same time, a reduction in cost was anticipated as inexpensive optical components such as aspheric lenses and high-speed digital signal processing electronics were being developed for consumer electronics. With hindsight, the "cost less and do more" design brief was a reasonable start point and it provided a well-defined envelope to work within.

Engineering design is inevitably a compromise between many often conflicting objectives and constraints. There are soft constraints such as cost, function, size and weight which might be flexible to a degree and hard constraints such as compliance with safety legislation and possible patent infringements that cannot be overlooked. For the case of a ceilometer, traditional designs, in which the receiver and transmitter telescopes are separate tubes, have been produced for many years and are not constrained by patents. In this arrangement, the transmitter and receiver are optically isolated but the instrument is less sensitive at low altitude where overlap is reduced. There are also coaxial ceilometer designs that exploit the same telescope for both receiver and transmitter but are constrained by multiple patents. In principle, coaxial designs offer better overlap characteristics, are potentially cheaper and more compact than the traditional design; however, although the receiver collects slightly more backscattered light from low altitude, optical isolation is reduced and greater sensitivity is difficult to achieve in practice.

In general, the process of engineering design is most rewarding when an idea is conceived to solve a particular problem and, when subsequently evaluated, several other advantages become apparent. Josh's thesis presents an innovative "split-lens" ceilometer that is one such design. It is traditional in the sense that it exploits separate transmitter and receiver telescopes but they use objective lenses that are manufactured by slicing a single aspheric lens into two semicircular components. With semicircular objectives, the transmitter and receiver telescopes can be placed in close proximity so that overlap can be maximised, while optical isolation is maintained. Although a semicircular aperture is somewhat unconventional, it is more uniformly filled by the elliptical beam of the semiconductor laser source, which means a proportionately more powerful laser can be used without additional beam shaping optics. It also means that the receiver and transmitter have matched optics of the same focal length, but with little additional cost.

Josh's thesis is more than an interesting design idea though. It covers mechanical design, alignment and describes how the backscatter signal is filtered, inverted and processed to find cloud layers and output vertical visibility. It also presents a new factory-based method to measure the instrument overlap that is necessary low altitude measurements of aerosols and pollutants.

The commercial ceilometer is, of course, an evolving product that is constantly being updated and improved by a team of dedicated engineers at Campbell Scientific Ltd. The team's contribution to Josh's work are acknowledged in the thesis but a special mention is due to James Richards who worked tirelessly designing, building and testing the digital signal processing electronics and software. Although I am sure that James has progressed somewhat the algorithms used in the commercial ceilometer from those described in the thesis, it is hoped that the content will be of interest to those wishing to understand ceilometers, what they do and why modern instruments are the way that they are.

Loughborough, October 2014 Prof. Jeremy Coupland

Abstract

A novel lidar ceilometer prototype based on divided lens optics has been designed, built, characterised and tested. The primary applications for this manufacturable ground-based sensor are the determination of cloud base height and the measurement of vertical visibility.

First, the design, which was developed in order to achieve superior performance at a low cost, is described in detail, along with the process used to develop it. The primary design considerations of optical signal-to-noise ratio, range-dependent overlap of the transmitter and receiver channels and manufacturability, were balanced to develop an instrument with good signal-to-noise ratio, fast turn-on of overlap for detection of close range returns, and a minimised number of optical components and simplicity of assembly for cost control purposes.

Second, a novel imaging method for characterisation of transmitter–receiver overlap as a function of range is described and applied to the instrument. The method is validated by an alternative experimental method and a geometric calculation that is specific to the unique geometry of the instrument. These techniques allow the calibration of close range detection sensitivity in order to acquire information prior to full overlap.

Finally, signal processing methods used to automate the detection process are described. A novel two-part cloud base detection algorithm has been developed, which combines extinction-derived visibility thresholds in the inverted cloud return signal with feature detection on the raw signal. In addition, standard approaches for determination of visibility based on an iterative far boundary inversion method, and calibration of attenuated backscatter profile using returns from a fully attenuating water cloud, have been applied to the prototype.

The prototype design, characterisation and signal processing have been shown to be appropriate for implementation into a commercial instrument. The work that has been carried out provides a platform upon which a wide range of further work can be built.

Acknowledgments

This work is the result of a joint project between Loughborough University and Campbell Scientific Ltd., part-funded by the UK Technology Strategy Board, to develop a novel laser-based atmospheric sensor. I would like to acknowledge Andrew Sandford, the technical director, and Dick Saffell, managing director at the time and now chairman of the board, for their strong support throughout the project of the academic side of the work. Many people at the company contributed in some way to the project, including David Foo who developed the first version signal processing algorithms. The instrument could not have been developed without the skill and creativity of James Richards, senior electronics engineer, with whom I worked on a daily basis throughout most of the project. I am grateful for the excellent guidance, enthusiasm and patience of Jeremy Coupland, my Ph.D. supervisor at Loughborough University. I am very fortunate to have had the opportunity to work with him. Finally, I must thank my family for all of the support and encouragement they gave me throughout my Ph.D.

Joshua D. Vande Hey

Contents

Nomenclature

(x_D, y_D)	Coordinates in the Detector Plane
(x_L, y_L)	Coordinates in the Laser Plane
(x_L, y_L)	Coordinates in the Receiver Lens Plane
(x_T, y_T)	Coordinates in the Object Plane
(x_W, y_W)	Coordinates in the Laser Lens Plane
$\alpha(r)$	Range-Dependent Atmospheric Extinction Coefficient
$\alpha_A(r)$	Range-Dependent Absorption Coefficient
α_L	Extinction at Lidar Wavelength
α_P	Extinction Observed by Pilot
α_{ave}	Average Extinction
$\alpha_{\min}$	Minimum Resolvable Extinction
$\beta'(r)$	Range-Dependent Attenuated Backscatter
$\beta'_{\text{mol}}(r)$	Calculated Molecular Attenuated Backscatter Profile
$\beta(r)$	Range-Dependent Backscatter Coefficient
β_m	Molecular Volume Backscatter Coefficient
$\beta_T(r)$	Range-Dependent Total Scattering Coefficient
$\Delta_{\lambda R}$	Transmitting Wavelength Bandwidth of Optical Filter
$\Delta_{\lambda T}$	Wavelength Bandwidth of the Laser
$\epsilon(r)$	Function Relating Scattering Plane Intensity to Intensity Incident on Receiver Objective
η	System Efficiency
η_A	Fraction of Laser Beam Transmitted by Objective Aperture
η_F	Transmission Efficiency of Optical Band-pass Filter
η_G	Transmission Efficiency of External Glass Plate
$\eta_m(r)$	Range-Dependent Multiple Scattering Correction Factor
η_O	Transmission Efficiency of Objective Lens
$\Gamma(x)$	Gamma Distribution Function
λ	Wavelength
$\mathcal{A}(r)$	Overlap Area
$\mathcal{F}$	Fourier Transform

$\mathcal{F}^{-1}$	Inverse Fourier Transform
$\mathcal{R}$	APD Responsivity
$\mathrm{L_a}$	Imaging Lens
$\mathrm{L_b}$	Relay Lens
μ	Term Describing Droplet Size Distribution Width
Ω	Field of View Solid Angle
ω	Frequency
ω_i	Frequency at Bin i
ω_L	Distance from Lidar Objective to Imaging Lens
ϕ	Phase Angle
Π_p	Constant Backscatter to Extinction Ratio
$\Pi_p(r)$	Range-Dependent Backscatter to Extinction Ratio
ρ	Ratio of Extinction at Lidar Wavelength to Extinction of Visible Light
$\sigma(t)$	Standard Deviation of Signal
σ_x	Gaussian Laser Distribution Constant along x-axis
σ_y	Gaussian Laser Distribution Constant along y-axis
τ	Laser Pulse Duration
τ_α	Optical Depth
θ_F	Full-angle Field of View
θ_H	Half-angle Field of View
θ_T	Maximum Half-Angle Transmitter Divergence
θ_V	Viewing Angle Below Horizon
$\Upsilon(x_T, y_T, r)$	Transmitted Laser Distribution Function
Υ_0	Laser Power Normalisation Constant
a	Droplet Radius
$a(\omega)$	Real Magnitude
A_0	Area of Receiver Objective
a_0	Mode of Droplet Distribution
a_o	Distance from Imaging Lens to Target
$A_R(x_L, y_L)$	Receiver Aperture Function
A_T	Area of Transmitter Objective
$A_T(x_W, y_W)$	Transmitter Aperture Function
A_{eff}	Effective Receiver Area
a_{eff}	Effective Droplet Radius
B	Integrated Attenuated Backscatter
B_0	Lidar Ratio Constant Multiplier
$B_1(\omega)$	Band-pass Filter 1
$B_2(\omega)$	Band-pass Filter 2
b_o	Distance from Target to Relay Lens
C	Laboratory Scattering and Transmission Constant
C_0	Actual Contrast
C_σ	Coefficient of Variance
C_D	Perceived Contrast

$C_{\beta'}$	Dimensional Attenuated Backscatter Normalisation Constant
D	Extinction Integration Distance
d	Laser Emitter Stripe Spacing
$D(x_D, y_D)$	Detector Aperture Function
d_D	Detector Diameter
d_L	Lens Diameter
E_{det}	Energy Collected by Photodetector
E_{obj}	Energy Incident on Receiver Objective
F	Outgoing Radiant Flux
f	Focal Length
$f(\lambda)$	Background Radiance at Wavelength λ
$f(a)$	Droplet Size Distribution Function
$f(T)$	Laser Lifetime Temperature-Dependent Multiplier
F_0	Incoming Radiant Flux
f_a	Focal Length of Imaging Lens
f_b	Focal Length of Relay Lens
F_R	Pulse Repetition Frequency
$G(r)$	Range-Dependent Geometric Factor
G_{A_1}	Gain of Primary Amplifier
G_{APD}	APD Gain
H	Thickness of a Layer of Transmitting Medium
$H(\omega)$	Frequency Domain Amplifier Impulse Response
$h(t)$	Amplifier Impulse Response
$I(\omega)$	Frequency Domain Amplifier Input
$i(t)$	Amplifier Input Signal
$I(x_D, y_D, r)$	Distribution in Focal Plane of Intensity Received
k	Lidar Ratio Exponent
K'	Luminance Contrast Threshold of Human Eye
K_L	Constant of Unknown Measurement Parameters
K_s	Lidar System Constant
L	Laser Emitter Length
l	Laser Emitter Stripe Length
$L(x_S, y_S)$	Near-Field Laser Output Distribution
L_D	Laser Power Density at Transmitter Objective
m	Real part of Complex Refractive Index
N	Molecular Number Density
N_B	Background Noise Counts
$N_B(r)$	Range-Dependent Background Noise
N_s	Molecular Number Density at Standard Temperature/Pressure
$O(\omega)$	Frequency Domain Amplifier Output
$O(r)$	Overlap Function
$o(t)$	Amplifier Output Signal

P	Signal Counts
$P(r)$	Range-Dependent Power
$P(t)$	Time Domain Signal
P_0	Average Laser Power During Pulse
P_L	Laser Power During Pulse
$P_M(r)$	Amplified Output Voltage Measured
$P_{B_1}(t)$	Return Signal Filtered using Band-pass Filter $B_1(\omega)$
$P_{B_2}(t)$	Return Signal Filtered using Band-pass Filter $B_2(\omega)$
$P_{D=\infty}(r)$	Power Incident on Infinite Aperture Detector
R	Range Bin Number
r	Range
$r(a_o)$	Perceived Range of Target from Lidar Objective
$r(b_o)$	Perceived Range of Target from Lidar Objective
r_0	Full Overlap Distance
r_b	Far Boundary Distance
R_c	Cloud Base Location
r_e	Pilot-Reported Cloud Base Height
r_f	Farthest Range at which SNR $\geq$ 6 dB
r_p	Distance from Target to Lidar Entrance Pupil
r_s	Selected Range Near Far Boundary
r_{base}	Range of Particle Layer Signal Onset
r_{peak}	Range of Particle Layer Signal Peak
R_{thr}	Particle Layer Significance Threshold
$S(r)$	Range-Dependent Extinction to Backscatter Ratio
S_c	Constant Ratio of Measured Returns to Modelled Molecular-only Returns
$S_c(r)$	Range-Dependent Ratio of Measured Returns to Modelled Molecular-only Returns
s_p	Distance from Imaging Lens to Lidar Entrance Pupil
S_t	Thin Cloud Return Peak Threshold
T	Temperature
$T(H)$	Optical Transmittance of a Layer of Thickness H
$T_{\text{tot}}(r)$	Total One-way Transmission along Path
u	Substituted Variable $u = \frac{r}{f} x_D$
v	Substituted Variable $v = \frac{r}{f} y_D$
V_d	Horizontal Visibility Threshold for Dense Cloud
$V_H(r)$	Horizontal Visibility as a Function of Vertical Range
V_i	Horizontal Visibility Calculated for Range Bin i
V_{MOR}	Meteorological Optical Range
V_{ave}	Average Horizontal Visibility Threshold
w	Laser Emitter Stripe Width
$w(r)$	Laser Beam Radius
x_0	Offset of Transmitter and Receiver Optical Axes
x_g	Offset of Laser Beam Centre from Transmitter Optical Axis

$Z(r)$	Range-Corrected Signal
a_o'	Distance from Imaging Lens to Image of Target Presented by Relay Lens
Ac	Altocumulus
APD	Avalanche Photodiode
As	Altostratus
BKN	Broken Cloud Layer
c	Speed of Light
Cb	Cumulonimbus
Cc	Cirrocumulus
CFARR	Chilbolton Facility for Atmospheric Research
Ci	Cirrus
CLR	Sky Clear
Cs	Cirrostratus
Cu	Cumulus
DIAL	Differential Absorption Lidar
F/#	F-number: focal length/diameter
FAA	Federal Aviation Administration
FEW	Few Clouds
FOTS	Fraction of Total Signal
FOV	Field of View
FWHM	Full Width at Half Maximum
GCM	General Circulation Model
HSRL	High Spectral Resolution Lidar
IR-corrected	Impulse Response-Corrected
ISO	International Organization for Standardization
MTTF	Mean Time to Failure
NRB	Normalised Relative Backscatter
Ns	Nimbostratus
OVC	Overcast
PM	Particulate Matter
RBC	Rotating Beam Ceilometer
Sc	Stratocumulus
SCT	Scattered Clouds
SKC	Sky Clear
SNR	Signal-to-Noise Ratio
St	Stratus
STRAT	Structure of the Atmosphere
TCu	Towering Cumulus
UTC	Coordinated Universal Time
WMO	World Meteorological Organization

Chapter 1
Introduction and Literature Review

Clouds and aerosols are part of everyday life around the planet, influencing weather, health, and climate. Local monitoring of clouds and visibility is important to air travel safety, local aerosol monitoring is an important part of air quality assessment, and global observations of clouds and aerosols are critical to understanding the radiative processes that drive climate change. According to the Intergovernmental Panel on Climate Change report in 2007 [1], the processes of both direct radiative forcing caused by scattering and absorption by anthropogenic aerosols and indirect radiative forcing caused by the influence of aerosols on cloud albedo and cloud lifetime are poorly understood, and while in general clouds and aerosols are expected to have overall cooling effects, there are large uncertainties on their total contributions to the Earth's radiation budget.

While clouds can be studied by satellite remote sensing [2], by balloon-borne radiosondes measuring temperature, pressure, and relative humidity in situ [3], or by in situ aircraft-based sensors such as nephelometers that measure the light-scattering properties of particles [4], ground-based passive and active remote sensing techniques generally offer advantages such as high temporal resolution and the capability to characterise part or all of a column of the atmosphere on a continuous or quasi-continuous basis with easy access to the instruments. Some useful passive sensors include sun photometers, which measure optical thickness of an atmospheric column by tracking power received from the sun at a number of wavelengths [5], infrared radiometers, which can measure cloud temperature by looking at its radiometric brightness in the 10–12 μm wavelength range [6], and microwave radiometers, which can measure temperature, water vapour content, and cloud liquid water content by monitoring radiometric brightness at a number of microwave frequencies [7].

Two active techniques are particularly important for atmospheric sensing. One of these is radar (radio detection and ranging), in which microwave signals are transmitted and backscattered radiation from scattering targets is measured, enabling the targets to be located by time of flight calculations. In addition, velocity of scatters can be determined by measuring frequency shift of the returned signals. Early meteorological applications for radar employed relatively long wavelengths, such as 3 cm [8], to detect precipitation or precipitating clouds, but more recent systems with smaller

J.D. Vande Hey, *A Novel Lidar Ceilometer*, Springer Theses,
DOI 10.1007/978-3-319-12613-5_1

wavelengths, such as 3 mm, can detect backscatter from particles as small as a few microns and can be used to retrieve cloud properties such as vertical distribution of liquid water [9]. The other major active technique for atmospheric sensing and the subject of this thesis is lidar (light detection and ranging). Lidar is similar to radar in that it utilises time-of-flight measurements, but it operates in a different region of the electromagnetic spectrum ranging from ultraviolet to infrared light, typically 250 nm–11 μm [10] and relies on lasers as transmitters. Because of the significantly shorter wavelengths involved, lidar can be used to detect much smaller atmospheric constituents, down to the molecular scale. In addition, a wide variety of spectroscopic measurements are possible. As an aside, note that sodar (sound detection and ranging) is another active technique related to radar and lidar in which acoustic waves (at a variety of angles) are transmitted and backscattered signals detected in order to determine range-resolved three-dimensional wind speeds at ranges of up to 1,000 m [11] through analysis of the horizontal and vertical Doppler frequency shifts from moving air.

Lidar ceilometers, also called laser ceilometers, are low-cost instruments that employ the lidar technique to determine height of cloud base and vertical visibility. Extended capabilities of these instruments include volcanic ash layer tracking [12] and boundary layer aerosol monitoring [13], which is becoming increasingly important as the Organization for Economic Cooperation and Development (OECD) is projecting major increases in health problems due to exposure to anthropogenic aerosols [14]. While ceilometers are at the low end of the lidar performance range, the fact that they are also on the low end of the cost range means that they can be deployed much more widely than more advanced lidar systems. This means that improving the performance of these instruments while making them straightforward to manufacture allows them to be used not just for aviation safety but for also for study of weather, air quality, and climate. They can therefore help to reduce uncertainty in the understanding of global cloud and aerosol processes and to improve understanding of public exposure to pollutants.

In this thesis, a new prototype lidar ceilometer capable of monitoring cloud base and sensitive to boundary layer aerosols is described. First, a novel divided-lens design that provides a solution to a classical lidar problem of balancing transmitter-receiver overlap and signal-to-noise ratio is introduced. The divided-lens approach maintains a high signal-to-noise ratio while optimising sensitivity in the near-range of the instrument in a design that is straightforward to manufacture and therefore has potential for broad deployment. Next, its optical characterisation by a new method is discussed. This technique allows the transmitter-receiver overlap of a lidar instrument to be profiled within a laboratory rather than over distances of hundreds of metres in the field. Finally, an original algorithm for automated cloud detection is described. Along with the description of the algorithm, preliminary prototype performance results are given by comparison with a higher specification research ceilometer. The research described here has since formed the basis for the development of a commercial cloud base sensor for meteorological applications, an instrument which is currently being refined for enhanced aerosol sensitivity.

1.1 Project Background and Objectives

The research for this thesis was funded jointly by Campbell Scientific, the UK Technology Strategy Board through their Knowledge Transfer Partnership Scheme, and Loughborough University. The objective of the project was to design, build, and characterise a lidar ceilometer prototype for the measurement of cloud base height and vertical visibility. It was intended that the prototype would be suitable for manufacture after a period of testing and refinement, and that the design of the instrument would offer a strong alternative to other products on the market. Prior to my involvement in the project, which began in 2007, an engineer by the name of Nicholas Cann spent half a year in collaboration with my supervisor Professor Jeremy Coupland laying some groundwork. Though he had not assembled a working instrument, Nicholas found and purchased suitable fundamental electronic and optical components for a test system, and most importantly, built two prototype rigs, one based on two circular aspheric lenses and one based on a compound divided-mirror design he had developed with Professor Coupland. This meant that when I began work on the project I could immediately begin putting pieces together to experiment with, and it also meant that I had two physical systems in front of me which would eventually inspire me to develop the divided aspheric lens design that has been used in the prototype.

1.2 Clouds and Aerosols in the Troposphere

1.2.1 The Troposphere

As source of most of the activity directly influencing weather on the surface of the Earth, the troposphere, the region of the atmosphere closest to the surface, is the layer of primary interest to meteorologists. The thickness of the troposphere is about 12 km on average, but can be 9 km or lower in polar regions, and greater than 16 km in the tropics [15].

1.2.2 Planetary Boundary Layer

The lowest 1–2 km of the troposphere is known as the planetary or atmospheric boundary layer. According to Kovalev and Eichinger [16], this region is the most intensely studied part of the atmosphere for a number of reasons. It is the source of the vast majority of the energy, water vapour, and chemical species distributed throughout the atmosphere. In addition, much atmospheric chemistry takes place here, and human activity, particularly emission of pollutants, has great influence.

Due to the combination of convective and turbulent flow in this region, surface emissions can be distributed throughout the layer quickly. As defined by Stull [17], the planetary boundary region is "the part of the troposphere that is directly influenced by the presence of the earth's surface, and responds to surface forcings with a time scale of about an hour or less." Kovalev and Eichinger further define the top of the boundary layer as the height that "is characterized by a sharp increase in temperature and a sudden drop in the concentration of water vapour and particulates as well as most trace chemical species [16]."

During the daytime in fair weather, a well-mixed boundary layer tends to form. This is referred to as a convective (or unstable) boundary layer. Stable boundary layers, on the other hand, occur when there is a temperature inversion in which temperature increases with height from the surface. This often happens at night or when dry air moves across wet surfaces. Kovalev and Eichinger caution that stable boundary layers can lead to dangerous air pollution events. If troublesome natural or human-created emissions at ground level are not dispersed well by mixing in the boundary layer, their concentrations at the surface increase. Therefore, monitoring of the boundary layer can provide information important for public health [16]. Note that boundary layer conditions are often complex and can be categorised more specifically than "stable" or "unstable" into a number of different types as has been done, for example, by Lock et al. [18].

1.3 Clouds Types, Structure, and Influence on Climate

The types, heights, and amounts of cloud have implications for both weather and climate. A brief summary of cloud types and structures and their roles in weather and climate is presented here. According to Ahrens [19], a general definition of a cloud is "a visible aggregate of tiny water droplets or ice crystals suspended in the air." The discussion of tropospheric clouds presented here in the following four sections is a standard approach to this subject. Here Ahrens [19] was used as a reference for a typical cloud classification system that distinguishes clouds of the most common types from each other and places them into four categories or families: high clouds, middle clouds, low clouds, and clouds with vertical development. Typical altitude ranges for the first three cloud categories are shown for different latitude zones in Table 1.1.

1.3.1 High Clouds

High clouds, in which ice crystals constitute the vast majority of water content, can generally be categorised as one of three types. Thin, wispy, heavily windblown, streamer like cirrus (Ci) are the most common among high clouds. They generally follow west to east prevailing winds and indicate fair weather.

Table 1.1 Approximate cloud base heights by latitude as described in [19]

Cloud group	Tropics (m)	Mid-latitudes (m)	Polar regions (m)
High clouds (Ci, Cs, Cc)	6,000–18,000	5,000–13,000	3,000–8,000
Middle clouds (As, Ac)	2,000–8,000	2,000–7,000	2,000–4,000
Low clouds (St, Sc, Ns)	0–2,000	0–2,000	0–2,000

Often appearing in expansive thin sheets, cirrostratus (Cs) can be so thin that the only indicator of their presence may be a refractive halo produced around the sun or the moon. Thick cirrostratus, which cast the sky glary white, often indicate advancing storm fronts and precede precipitation by 12–24 h, particularly if followed by middle.

The third and final type of high cloud is the cotton-puff-like cirrocumulus (Cc). These can occur individually or in woven rows, and they can produce striking visual effects at sunset by enhancing scattering of yellow or red light.

In addition, there is a fourth type of high cloud called sub-visual cirrus. Lynch [20] specifies sub-visual cirrus as cirrus clouds with maximum optical depths 0.03 (optical depth is discussed in Chap. 2), but points out that this commonly accepted limit is not physically-motivated. According to Lynch these clouds are often present at the tropopause (the boundary between the troposphere and the stratosphere above) and may be nearly omnipresent in tropical regions. Lynch also classifies cirrus contrails, the mixing clouds created by jet aircraft, as a fifth type of high cloud.

1.3.2 Middle Clouds

Middle clouds can generally be placed into one of two categories. The first of these are the grey or blue-grey altostratus (As) which often extend across the sky in large but somewhat diffuse-looking sheets. Made up of a mix of ice crystals and water droplets, they often semi-obscure the sun or moon which then appears as a watery disk. They can be distinguished from cirrostratus by their darker colour and lower height; the facts that halos are formed only by cirrostratus and that shadows are not generally visible when altostratus are present can also be used to tell the two types apart. Altostratus often indicate storms to follow with fairly continuous and widely distributed precipitation.

The puffy grey altocumulus (Ac) is the second type of middle cloud. Typically less than 1 km in vertical extent and consisting primarily of water droplets, these clouds sometimes appear in arrays of cloud puffs or puffy hay-row type structures. They generally appear darker in some parts of the cloud than in others, which helps to distinguish them from cirrocumulus, and they generally appear rounded or rolled, which helps to distinguish them from altostratus. Castle-like altocumulus castellanus indicate rising air in the cloud layer, and the appearance of these clouds in warm, humid morning weather suggests the possibility of afternoon thunderstorms.

1.3.3 Low Clouds

Low clouds, whose bases are below 2,000 m, are generally assigned to one of three different types. The first of these is the dark grey nimbostratus (Ns) which generally produce light or moderate but steady precipitation. The bases of these clouds appear diffuse and are difficult to discern, and their vertical extent can exceed 3 km. Below a nimbostratus layer, evaporating rain mixing with the air often leads to poor visibility and fog or a ragged, broken lower cloud layer. Nimbostratus can be distinguished from altostratus by their generally darker appearance and the fact that typically they completely obscure view of the sun or moon.

The second type of low cloud is the lumpy, tufted stratocumulus (Sc), light to dark grey clouds that can appear in rows, patches, or as more greatly-spaced elements. They often develop late in the day as a larger cumulus breaks up. Occasionally they produce showers in winter if they increase in vertical extent and their tops cool to −5 °C. Stratocumulus can be distinguished from altocumulus by the larger appearance of the cloud elements.

The third type of low cloud is the uniform, grey stratus (St) that has the appearance of a fog suspended above the surface and is often seen to obscure the whole sky. Frequently present in coastal areas in the summer, these clouds can form after a heavy fog begins to diminish, and though they do not generally produce precipitation, they may occasionally produce light mist or drizzle. A stratus layer can be distinguished from nimbostratus by its lower, often more uniform base and its lack of precipitation; stratus can be distinguished from altostratus by its lower height, darker grey appearance and greater obscuration of the sun and moon.

1.3.4 Clouds with Vertical Development

Clouds with vertical development are of great interest to meteorologists because these can become storm clouds. The white cotton-puff-like cumulus (Cu) usually have sharp outlines and flat, white or light grey bases. In humid conditions they can be as low as 1 km above the ground and approximately 1 km wide. The rounded tops of cumulus clouds are not particularly high, and the individual clouds in a layer are well-spaced. The rounded tops and greater spacing between clouds helps to distinguish them from the flatter, more closely-grouped stratocumulus. Fair weather cumulus (cumulus humilis) have limited vertical development.

Fair weather cumulus formed early on warm summer days can develop vertically as the day progresses. If a cumulus cloud has grown in height and begins to appear cauliflower-like with a dark base, it is categorised as a towering cumulus (TCu), also known as cumulus congestus. It may join with other towering cumulus to form a long line of cloud, and it may emit showery precipitation.

Towering cumulus can develop further vertically to become immense, cumulonimbus (Cb) storm clouds. The bases of these clouds can be as low as 600 m, and the clouds can extend more than 12 km to reach all the way to the tropopause. They may occur as individual clouds, or they may be incorporated into a wall of cloud. The warmer, lower parts of cumulonimbus consist of water droplets alone. In the middle region of the cloud both water droplets and ice crystals can be found, while at the cold cloud top, only ice crystals are present. High winds at the top of the cloud can reform the upper part to give it a flattened anvil appearance (cumulonimbus incus).

Condensation of water vapour in cumulonimbus releases great amounts of energy which leads to powerful updrafts and downdrafts that can exceed wind speeds of 36 m/s. Lightning, thunder, all types of heavy precipitation, and sometimes even tornadoes can occur with these clouds. In contrast to towering cumulus, which have clearly defined cloud tops, cumulonimbus have less well-defined, fibrous-looking tops. The presence of lightning, thunder, and large hail also distinguishes cumulonimbus from towering cumulus. Sketches of the the major cloud types in each of the four groups are shown in Fig. 1.1.

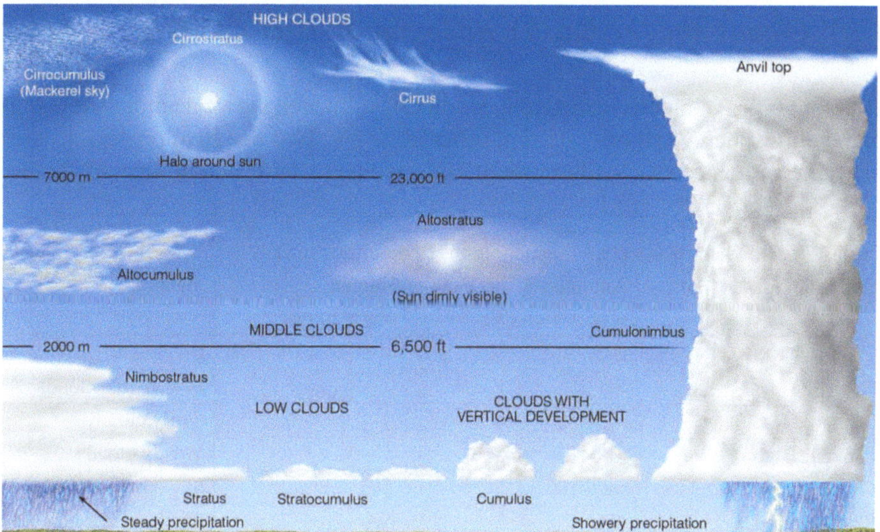

Fig. 1.1 Major cloud types, reproduced from [19] (used with permission)

1.3.5 Alternative Cloud Categorisation Approach

Note that according to Lynch [20] the cloud categorisation scheme that has been presented here based on Ahrens [19] aligns with the World Meteorological Organisation's (WMO) categorisation by height. However, he suggests that since categorisation by height groups clouds of different composition together (for example water droplet and ice crystal mixed altostratus and water droplet-based altocumulus) a more logical approach might be to categorise based on water phase content, since this would place clouds into structural groups and could help unify discussion among different disciplines such as planetary physics, crystallography, and remote sensing.

1.3.6 Sky Conditions

The amount of cloud cover is typically reported as a sky condition. Table 1.2 shows the standard sky conditions categories and criteria for both automated and manual observations. Note that cloud cover amount for automated observations is defined in terms of percentages, while cloud cover amount for human observations is estimated in oktas, or eighths of sky cover. Although automated observations have widely replaced visual observations, this sky condition categorisation structure is still widely used and ceilometers are expected to report it.

1.3.7 Clouds and the Earth's Radiation Budget

Different types of clouds in different conditions influence the total amount of energy in the atmosphere-surface system in different ways. Two radiative processes determine the influence of clouds in the earth's radiation budget. The first of these is

Table 1.2 Cloud cover sky condition definitions as described in [19]

Sky condition	Automated observation (%)	Human observation	Meaning
Clear (CLR/SKC)	0–5	0	No clouds
Few (FEW)	>5–25	$0 - \frac{2}{8}$	Few visible clouds
Scattered (SCT)	>25–50	$\frac{3}{8} - \frac{4}{8}$	Partly cloudy
Broken (BKN)	>50–87	$\frac{5}{8} - \frac{7}{8}$	Mostly cloudy
Overcast (OVC)	>87–100	$\frac{8}{8}$	Sky covered by clouds
Sky obscured	NA	NA	Sky hidden by surface-based fog, blowing snow, smoke, etc.

the albedo effect of reflecting shortwave (primarily visible or near-visible) incoming solar radiation and therefore reducing the amount of solar energy passing through them. Cloud albedo is poorly understood because it depends non-linearly on both concentration of cloud condensation nuclei and column-integrated liquid water content [21]. The second radiative process is the greenhouse effect whereby clouds prevent the escape of energy through the atmosphere into space by absorption and emission of longwave infrared radiation.

According to Ramanathan et al. [22], the net global radiative impact of clouds is cooling of 13.2 W/m^2. However, cirrus clouds, depending on their composition can either trap outgoing infrared radiation from the earth by the greenhouse effect or reflect incoming solar radiation by the albedo effect [23]. According to Del Genio, the current lack of knowledge of cirrus clouds and the processes that create them requires that large assumptions be made in general circulation model (GCM) simulations. This, in turn, leads to significant uncertainty in their role in a changing climate, to the extent that it is not clear whether they feedback positively or negatively in response to changing temperatures [24].

1.4 Aerosols

Aerosols are gaseous suspensions of solid and liquid particles [25]. In common practise in the atmospheric sciences, suspended particle types considered to be aerosols include all liquid and solid particles with the exception of hydrometers such as cloud droplets, raindrops, and ice crystals. Aerosols in the atmosphere range from a few molecules in size to greater than 100 μm in diameter. Two aerosol categorisation schemes described by Pruppacher and Klett [25] are shown in Table 1.3.

The difference between fine (less than approximately 1 μm) and coarse (greater than approximately 1 μm) aerosols is significant [26]. According to Willeke and Whitby [27], fine aerosols are formed primarily by condensation. Transient, nuclei mode aerosols are often observed as fresh combustion aerosols. Accumulation mode aerosols arise through growth of smaller aerosols by coagulation of nuclei aerosols, or by condensation or combustion, and tend to remain in the atmosphere for days. On the other hand, coarse aerosols, such as dust and sea spray, are generated primarily by mechanical processes. Though they tend to settle fairly quickly, their suspension times in the atmosphere vary considerably.

Table 1.3 Aerosol size categorisation schemes as described in [25]

Junge		Whitby		
Dry radius r (μm)	Particle class	Diameter d (μm)	Particle mode	
$r < 0.1$	Aitken	$d < 0.1$	Nuclei	Fine
$0.1 \leq r \leq 1.0$	Large	$0.1 \leq d <\sim 1$	Accumulation	Fine
$r > 1.0$	Giant	$d >\sim 1$	Coarse	

According to Pöschl [28], particulate matter (PM) in the troposphere is made up primarily of sulphate, nitrate, ammonium, sea salt, mineral dust, organic compounds, and black (elemental) carbon, or soot. Each of these constitutes in the range of 10–30 % of the total by mass. With time in the atmosphere aerosols evolve through dissolution, condensation into larger particles, and chemical reactions with atmospheric gases. Particles with diameters $\leq 1\,\mu m$ are classed as PM1, those with diameters $\leq 2.5\,\mu m$ as PM2.5, and those with diameters $\leq 10\,\mu m$ as PM10. Most dangerous to human health are ultrafine particles with diameters of less than $0.1\,\mu m$, since these can be absorbed by the lungs directly into the blood stream. However, Pöschl emphasises that exactly which aerosol properties, such as size, mass, structure, solubility, etc., are most important in terms of health impact is not well understood.

Aerosol particles can act as nucleation sites for cloud particles. Sub-micron sulphate aerosols, which are generated both by natural processes such as phytoplankton dimethyl sulphide production and human industrial processes, are a major type of cloud condensation nuclei [21]. Bréon et al. found that cloud droplet size tends to be significantly smaller over highly-polluted continental areas, than in unpolluted tropical ocean regions [29]. Due to a larger number of aerosols to act as cloud condensation nuclei in polluted regions, the number of droplets increases and the mean droplet size decreases. Though it is not fully clear which aerosols are anthropogenic and which are natural, Brëon et al. explain that highest aerosol outputs are typically found in slash-and-burn agricultural areas and polluted urban areas. Aerosols themselves on average lead to global cooling by increasing the Earth's albedo. By increasing the number of cloud particles and decreasing their size, aerosols may also indirectly increase cloud albedo through a process called the Twomey effect [30]. According to Feingold et al., however, this effect can have either positive or negative radiative impact depending upon conditions [31].

1.5 Motivation for Measurement of Clouds and Aerosols

Clouds are important to human activity in a number of ways. They indicate weather conditions, produce precipitation, and generate phenomena such as lightning and tornadoes. They can obscure pilot visibility, and convective clouds are dangerous for aircraft. Clouds can reflect incoming sunlight and/or trap surface-emitted infrared radiation. Cloud measurement is of great importance to climate study, weather prediction, and air transportation safety. Since clouds are indicators of the state of the local atmosphere, knowledge of the height, type, and extent of cloud layers provides insight into the conditions and processes taking place in different parts of the atmosphere.

Aerosols, too, require study by remote sensing methods. Monitoring of aerosols is particularly important for public health. In addition, aerosols have a net cooling effect on the climate and also act as cloud nucleation sites which further influences climate. Therefore measurements leading to better understanding of aerosol generation, transport, removal and chemistry, and their interactions with clouds are necessary for better air quality forecasting and a more thorough understanding of radiative forcing processes in the atmosphere.

1.6 Lidar Remote Sensing of the Atmosphere

The earliest lidar measurements of the atmosphere were reported in 1963 by Ligda [32] and Fiocco and Smullin [33] and in 1964 by Fiocco and Grams [34], just a few short years after Maiman's demonstration of the first working laser in 1960 [35]. Lidar measurements have broad application in the characterisation of the atmosphere, ranging from the determination of properties of cloud particles [36] or aerosols [37], to the profiling of trace gas concentrations [38], air temperature [39], or wind velocity [40].

1.6.1 Lidar System Configurations

A typical lidar consists of two subsystems, a laser transmitter and a receiver. The laser transmitter may consist simply of a laser source, or it may include additional components such as beam-expanding optics [10]. The receiver generally consists of a telescope with a detector placed at, or sometimes slightly offset from [41], its focal plane. Lidar instruments can be arranged in a number of different ways as will be discussed later but the primary configuration types are biaxial, in which the laser transmitter is located beside the receiver, sometimes tilted slightly toward it [42], and coaxial, in which the laser beam shares the receiver's optical axis through implementation of mirrors and/or beam splitters [43].

1.6.2 Elastic Backscatter Lidar

One of the tools used for the study of clouds and aerosols is elastic backscatter lidar. Elastic scattering of light, discussed in more detail Chap. 2, is the interaction of light with scattering media whereby light at a given wavelength is scattered without alteration of the atomic or molecular energy states of the scatterers. For molecules or very small particles the scattering process can be described by Rayleigh scattering theory [16] and for particles of sizes similar to or larger than the wavelength of the light being scattered, the process can be described by Mie theory [44]. Elastic backscatter lidars measure light backscattered by elastic processes. Though spectroscopic absorption and/or excitation processes may be present, as well as wavelength shifts due to Doppler broadening by moving scatterers, the instrument is blind to these effects, and the contributions of these processes to an elastic channel are generally disregarded with the possible exception of absorption by well-understood concentrations of atmospheric constituents at the laser wavelength. Broadband absorption, however, is an important consideration for all lidar systems.

Despite the lack of spectroscopic resolution, there is still a wealth of information that can be obtained through elastic lidar measurements of the atmosphere, particularly if some additional information, such as extinction somewhere along the profile, is known during the time of the measurement. Typically, an exponential decrease in returns with range reveals the presence of the pure molecular atmosphere, and deviations from this exponential profile indicate the presence of more strongly-scattering aerosols and cloud droplets at ranges corresponding to these deviations [45].

Elastic-only lidars can be used effectively to determine concentrations of particles in the atmosphere if the scattering properties of particles in the measurement volume are known [46]. They can be used to retrieve cloud base height [47], estimate visibility [48], and with varied receiver fields of view, can be used to determine droplet sizes [49]. Based on the shape of the return signature they can also be used to discern precipitation type [50]. For regions of the atmosphere free of clouds and aerosols, backscatter intensity provides an indication of density, which can in turn be used to establish temperature [51]. Elastic lidars are the simplest type of lidar, but their signals tend to be the most difficult to invert because two parameters, extinction and backscatter, often need to be derived from one measurement, the intensity of backscattered light detected.

1.6.3 Depolarisation Lidar

One important extended technique that can be applied to elastic lidar systems is resolution of depolarisation. The orientation of the orthogonal electric and magnetic components of propagating light waves determine its polarisation. While spherical droplets do not significantly alter the polarisation of light they scatter, non-spherical particles such as ice crystals do alter the polarisation [52]. If the laser light transmitted from an elastic lidar system is highly polarised and the receiver is equipped with a means (such as a polarising beam splitter) of resolving the polarisation of returning light, the degree of depolarisation of backscattered light can be determined, and some understanding of the shapes of the scattering particles can be gained. Depolarisation lidar, discussed in detail by Sassen [53], resolves the polarisation ratio of the light backscattered from a volume illuminated by a polarised laser. Specifically, it determines the extent to which the polarisation of light is shifted away from the polarisation of the laser, as clarified by Gimmestad [54]. Application of the depolarisation technique can provide important information about the shapes of scatterers in the atmosphere. Furthermore, if the relationship between backscatter and extinction in the measurement volume is known, this technique can be used to make more subtle distinctions such as distinguishing ice from dust [55].

1.7 Commercial Ceilometers

The first patent for a laser ceilometer was filed in 1974 by Segre and Truscott [56]. After gradually replacing the searchlight-based rotating beam ceilometers of the past [57], commercial eye-safe lidar systems capable of measuring height of cloud base and reporting vertical visibility have become standard atmospheric measurement tools. Used widely for aviation and meteorology, these instruments typically transmit short, low power, near-infrared laser pulses at high repetition rates in order to adhere to eye-safety guidelines. A wavelength of 905 nm is commonly used for a number of reasons. First of all, there is reasonably good atmospheric transmission of about 0.6 at 905 nm. Secondly, inexpensive laser diodes with sufficient output power at 905 nm and silicon avalanche photodiodes (APDs) with good responsivity at this wavelength make instruments based on these components affordable. Finally, aviation instrumentation rules require that laser light transmitted upward from sensors located near airfields be invisible in order to avoid interfering with pilots' vision.

1.7.1 Belfort

One of the first commercially-available lidar ceilometers was produced by the USA-based Belfort Instrument. The Belfort model 7013 C was capable of detecting clouds at heights from 15 to 7,350 m [58]. Whereas the other instruments discussed here employ lenses, this one used a biaxial configuration based on two large mirrors. While these instruments were large and heavy and have now been largely retired, they were extremely rugged.

1.7.2 Eliasson

An instrument that has been deployed especially for aviation applications is the CBME80, manufactured in Sweden by Eliasson Engineering, AB. According to its specification, this biaxial instrument detects clouds ranging from 10 to 7,500 m above the surface [59]. While this is a fairly standard instrument in terms of performance, it offers the advantage of having a light weight of 15 kg.

1.7.3 AWI

The USA-based AWI (All Weather, Inc.) offers two biaxial ceilometer models. Their base model, the 8339, has a specified range of 7,600 m, while their higher-end model, the 8340, has a specified range of 12,200 m [60]. While these have found widespread use in North America, particularly for aviation applications, these instruments, somewhat surprisingly, do not appear to have been utilised for any study in the literature despite their impressive range.

1.7.4 Vaisala

Vaisala, Oyj., based in Finland, has produced two standard ceilometer models. Their current model, the Vaisala CL31 [61] has a specified altitude range of 7,500 m [62]. This instrument, which has a coaxial, common optics design where the laser beam and receiver utilise different regions of the same lens, replaced a previous model with a similar specification, the CT25K [50], which had fully-shared common optics. Both of these instruments are based on 905 nm laser diode sources. One advantage of both the CT25K and the CL31 over the biaxial instruments of other manufacturers is good sensitivity in the near-range. These instruments have been used for research, for example, to study cloud base height [63] and to study boundary layer aerosols [64, 65]. More details of the CL31 are given in Chap. 2.

Vaisala has offered a number of "research-grade" ceilometers at various times. One of these was a biaxial instrument called the LD-40, which had used a wavelength of 855 nm and had a specified range of 13,000 m [50]. Another was the CT75k, which was constructed out of four CT25k instruments bundled together [66] and had a maximum working range of 11.25 km [67]. A CT75k located at Chilbolton Observatory has been calibrated [68] and was used in this study as a reference for prototype measurements as described in Chap. 5. Vaisala's current research ceilometer model is the CL51, which also utilises a single-lens design, has a range of 13 km for cloud detection and 15 km for backscatter profiling [69].

1.7.5 Jenoptik

Another advanced ceilometer is offered by Jenoptik, GmbH, based in Germany. Their CHM 15 k family of instruments has a specified range of 5 m to 15 km and is designed to have good sensitivity to either aerosols and cirrus clouds depending upon the model. Unlike other ceilometers, which are based on laser diodes typically at 905 nm, this instrument utilises an Nd:YAG solid state laser at 1,064 nm. It is this laser which gives the CHM 15 k, the cirrus model, its higher range, but also adds significantly to the cost of the instrument. A network of CHM 15 k instruments installed by the German Weather Service (DWD) has been used to study aerosol layer thickness [70]. This network has also been utilised to track a volcanic ash layer [12]. A slightly different model, the CHM 15kx, has a wider field of view for improved aerosol profiling, but is not as sensitive to cirrus clouds as the CHM 15 k. Recently an aerosol retrieval method for this instrument has been described [71].

1.7.6 Leosphere

Leosphere, based in France, has recently introduced the R-Man 510 Super Ceilometer. This instrument is based on an Nd:YAG laser and emits 355 nm light. In addition to measuring elastic backscatter intensity, this instrument can provide information about scatterer shapes by the use of a depolarisation channel. It also includes a nitrogen Raman channel which is used to provide molecular signal calibration necessary for an accurate inversion of the signal. The range of the instrument is specified as 4.5 m–20 km and it has been designed specifically with the goal of measuring volcanic ash concentrations accurately for aviation safety applications [72].

1.7.7 Other Ceilometers

A few other instruments, less well known, are also available. The first of these is the ALC30 from Degreane Horizon in France. This instrument is based on an Erbium-doped glass laser at the eye-safe wavelength of 1,535 nm, but has a range of only 15–7,500 m [73]. The second is from MTECH Systems in Australia. Their 905 nm diode-based 8200-CHS has a specified range of up to 8,000 m [74]. Finally, there is an extremely small, hand-held ceilometer for military applications available from Tempestini in Italy. Their TMP-09-03 is based on an erbium-doped glass laser at 1.54 μm [75]. It is extremely portable but does not specify a range.

1.7.8 Prototypes in the Literature

Much of the work in the field of ceilometer design has been conducted in industry, so many of the subtleties of instrument design that have been learned are kept as trade secrets. However, a 905 nm prototype lidar ceilometer design was developed by Gregorio et al. at Universitat Politcnica de Catalunya in Spain [76, 77]. First, they modelled and measured the signal to noise ratio for various receiver field of view (FOV) angles and receiver objective areas in a biaxial lidar [76]. Then, they presented an inexpensive ceilometer design using a Fresnel lens for the receiver [77]. Their work focused on optimizing design parameters of receiver diameter, FOV, and laser tilt angle using low cost laser diodes and APD detectors. This work was recently expanded to include more detail [42].

A significant attribute of their design was a slit-shaped field stop. In a biaxial lidar system where the transmitter is located next to the receiver, defocus arises because at short ranges, the backscattered signal returns will not be centred and focused at the focal point of the receiver lens, but rather off to the side and out of focus. Because defocus is a major contributor to the overlap function, the placement and shape of the aperture are significant in managing the dynamic range and low altitude performance of a biaxial lidar system [41]. The slit-shaped aperture employed by Gregorio et al.

[77] allowed for increased detection of off-axis returns by increasing the FOV along the slit without making the entire FOV larger, thus enhancing close-range overlap while largely suppressing the corresponding increase in daytime background noise typically resulting from a larger FOV.

1.7.9 Atmospheric Research Utilising Ceilometers

Beyond the monitoring of cloud base height and vertical visibility, ceilometers have been used for a number of research applications including the study of signatures of different types of precipitation [50]. The most important extended application, however, is the retrieval of aerosol properties. Ceilometers often lack the stability and sensitivity of more refined instruments and typically have one elastic channel only. For these reasons, the aerosol-related parameter most realistic for study using ceilometers is the height of the planetary boundary layer.

As the planetary boundary layer contains the majority of the atmosphere's aerosols, which show significantly stronger returns than do the molecular atmosphere, its height can be found by locating the first significant decrease in lidar return intensity [65]. Note, however, that "boundary layer height" reported by this method is actually the height of the top of the aerosol layer closest to the surface. Therefore, higher aerosol layers and more complex boundary layer structures are disregarded and the "boundary layer height" may or may not relate directly to temperature inversion. Nevertheless, a number of authors, including Münkel and Räsänen [61] and Wiegner et al. [78], have considered the application of this type of technique.

Wavelet filtering techniques, which locate features in a signal through the use of filters based on families of wavelet functions [79], have also been applied to identify the top of the boundary layer from lidar signals [80]. These have been applied to ceilometers by, for example, Teschke et al. [81] and de Haij et al. [82]. Recently a study by Haeffelin et al. [83] found little difference in the ability of various derivative-based and wavelet-based techniques to identify aerosol layer gradients. The same study showed, however, that it is difficult to distinguish the top of the nighttime mixing layer from the top of the residual layer by an elastic lidar gradient method. The ability of the lidar measurements to locate the boundary layer height within 300 m of that determined by radiosonde dropped from 67 % during the day time to 33 % at night. Haeffelin et al. also found that when near-infrared ceilometer data and UV lidar data both indicated the same mixing layer height, this value was 25–40 % more likely to match within 300 m that determined by radiosonde than when either the lidar or the ceilometer was used alone.

Boundary layer height has also been determined by assimilation of measured data with modelled data. Di Giuseppe et al. [84] described a method wherein the signal discontinuities from individual ceilometer measurements, which indicate possible aerosol layer structures, are coupled with a time-dependent model of the boundary layer in order to identify the most statistically significant feature as the top of the boundary layer. While this method is more involved and requires a timescale of one day for processing, its boundary layer height assignments agree more closely with radiosonde measurements than do those derived from ceilometer measurements alone, particularly at night when residual aerosol layers often exist above the mixing layer. In some cases this approach was demonstrated to reduce to tens of metres ceilometer-based boundary layer measurement errors that would have otherwise been on the order of 1,500 m. It is therefore a powerful technique for enhancing the accuracy of ceilometer-reported boundary height.

In other work on ceilometer aerosol retrievals, Münkel et al. [85] compared ceilometer, radar, sodar, and in situ particulate measurements of the urban boundary layer. They found that PM10 (10 μm particulates) concentrations showed an essentially linear relationship to backscatter intensity measured by the lidar. This suggests that in constrained conditions particulate pollutant concentrations might be inferred from ceilometer data. In related work, Heese et al. [86] have considered the effectiveness of aerosol optical property retrieval by ceilometers, Tsaknakis et al. [65] performed a comprehensive study of urban boundary layer involving lidar, ceilometer, and radiosonde, and Stachlewska et al. [87] have devised a novel variable spatial and temporal averaging technique to optimise the signal to noise ratio of ceilometer returns from boundary layer aerosols and extract transient features such as cloud layers. While some important work has been done in this area, more study of the most efficient and effective means of exploiting ceilometer data for aerosol retrievals is required.

1.8 Overview of Thesis

Following this introduction, Chap. 2 summarises elastic lidar theory relevant to the work in this thesis, placing emphasis on the lidar equation and its inversion. Chapter 3 presents the opto-mechanical design of a novel lidar ceilometer prototype suitable for manufacture and the design approach that was applied in order to develop it. Chapter 4 describes a theoretical method for the calculation of the overlap of a lidar transmitter and receiver throughout its range and presents and evaluates a novel imaging-based laboratory measurement of overlap. Chapter 5 presents a novel two-part signal processing algorithm that has been developed for automated determination of cloud base height and demonstrates its performance, and also discusses lidar calibration and determination of vertical visibility from lidar returns. Finally, Chap. 6 summarises the conclusions that can be drawn from this research and discusses further work emerging from it.

References

1. S. Solomon, D. Qin, M. Manning, Z. Chen, M. Marquis, K. Averyt, M. Tignor, H.L. Miller (eds.), *Contribution of Working Group I to the Fourth Assessment Report of the Intergovernmental Panel on Climate Change* (Cambridge University Press, Cambridge, 2007)

2. G.L. Stephens, D.G. Vane, R.J. Boain, G.G. Mace, K. Sassen, Z. Wang, A.J. Illingworth, E.J. O'Connor, W.B. Rossow, S.L. Durden, S.D. Miller, R.T. Austin, A. Benedetti, C. Mitrescu, T. CloudSat Science Team, The CloudSat mission and the A-Train. Bull. Am. Meteorol. Soc. **83**(12), 1771–1790 (2002)
3. C.M. Naud, J.-P. Muller, E.E. Clothiaux, Comparison between active sensor and radiosonde cloud boundaries over the ARM southern great plains site. J. Geophys. Res. **108**(D4), 4140 (2003)
4. J.F. Gayet, O. Crpel, J.F. Fournol, S. Oshchepkov, A new airborne polar nephelometer for the measurements of optical and microphysical cloud properties. Part I: Theoretical design. Ann. Geophys. **15**, 451–459 (1997). doi:10.1007/s00585-997-0451-1
5. M. Morys, F.M. Mims, S. Hagerup, S.E. Anderson, A. Baker, J. Kia, T. Walkup, Design, calibration, and performance of MICROTOPS II handheld ozone monitor and sun photometer. J. Geophys. Res. **106**(D13), 14573–14582 (2001)
6. S.Y. Matrosov, T. Uttal, J. Snider, R. Kropfli, Estimation of ice cloud parameters from glround-based infrared radiometer and radar measurements. J. Geophys. Res. **97**(D11), 11567–11574 (1992)
7. J. Liljegren, E. Clothiaux, S. Kato, B. Lesht, Initial evaluation of profiles of temperature, water vapor and cloud liquid water from a new microwave profiling radiometer, in *Proceedings of 5th American Meteorological Society (AMS) Symposium on Integrated Observing Systems*, Albuquerque, NM, 14–19 Jan 2001
8. J.S. Marshall, Precipitation trajectories and patterns. J. Meteorol. **10**(1), 25–29 (1953)
9. E.E. Clothiaux, M.A. Miller, B.A. Albrecht, T.P. Ackerman, J. Verlinde, D.M. Babb, R.M. Peters, W.J. Syrett, An evaluation of a 94-GHz radar for remote sensing of cloud properties. J. Atmos. Oceanic Technol. **12**(2), 201–229 (1995)
10. U. Wandinger, Introduction to lidar, in *LIDAR: Range-Resolved Optical Remote Sensing of the Atmosphere*, ed. by C. Weitkamp (Springer, Berlin, 2005), pp. 1–18
11. K. Baumann, M. Piringer, Two-years of boundary layer measurements with a sodar—statistics and applications. Phys. Chem. Earth Part B Hydrol. Oceans Atmos. **26**(3), 205–211 (2001)
12. H. Flentje, H. Claude, T. Elste, S. Gilge, U. Köhler, C. Plass-Dülmer, W. Steinbrecht, W. Thomas, A. Werner, W. Fricke, The Eyjafjallajökull eruption in April 2010-detection of volcanic plume using in-situ measurements, ozone sondes and lidar-ceilometer profiles. Atmos. Chem. Phys. **10**(20), 10085–10092 (2010)
13. I. Stachlewska, S. Migacz, A. Szkop, A. Zielinska, M. Piadlowski, P. Swaczyna, K. Markowicz, S. Malinowski, A. Gorska, Annual variability of atmospheric boundary layer in Warsaw, in *Reviewed and Revised Papers of 26th ILRC International Laser Radar Conference*, 25–29 July 2012, Porto Heli, Greece, pp. 945–948
14. OECD, OECD Environmental outlook to 2050 (OECD Publishing, 2012), http://dx.doi.org/10.1787/9789264122246-en
15. F. Lutgens, E. Tarbuck, D. Tasa, *The Atmosphere: An Introduction to Meteorology*, 9th edn. (Prentice Hall, Upper Saddle River, 2004)
16. V. Kovalev, W. Eichinger, *Elastic Lidar: Theory, Practice, and Analysis Methods* (Wiley, Hoboken, 2004)
17. R. Stull, *An Introduction to Boundary Layer Meteorology*, Atmospheric Sciences Library (Kluwer Academic Publishers, Boston, 1988)
18. A.P. Lock, A.R. Brown, M.R. Bush, G.M. Martin, R.N.B. Smith, A new boundary layer mixing scheme. Part I: scheme description and single-column model tests. Mon. Weather Rev. **128**(9), 3187–3199 (2000)
19. D.C. Ahrens, *Meteorology Today: An Introduction to Weather, Climate, and the Environment*, 9th edn. (Brooks/Cole, Belmont, 2009)
20. D.K. Lynch, Cirrus, history and definition, in *Cirrus* (Oxford University Press, New York, 2002), pp. 3–10
21. P.G. Falkowski, Y. Kim, Z. Kolber, C. Wilson, C. Wirick, R. Cess, Natural versus anthropogenic factors affecting low-level cloud albedo over the North Atlantic. Science **256**(5061), 1311–1313 (1992)

22. V. Ramanathan, R.D. Cess, E.F. Harrison, P. Minnis, B.R. Barkstrom, E. Ahmad, D. Hartmann, Cloud-radiative forcing and climate: results from the earth radiation budget experiment. Science **243**(4887), 57–63 (1989)
23. K. Sassen, cirrus, a modern perspective, in *Cirrus* (Oxford University Press, New York, 2002), pp. 11–40
24. A.D. DelGenio, GCM simulations of cirrus for climate studies, in *Cirrus* (Oxford University Press, New York, 2002), pp. 310–326
25. H. Pruppacher, J. Klett, *Microphysics of Clouds and Precipitation* (Kluwer Academic Publishers, Dordrecht, 1997)
26. K.T. Whitby, The physical characteristics of sulfur aerosols. Atmos. Environ. **1967**(12), 135–159 (1978)
27. K. Willeke, K.T. Whitby, Atmospheric aerosols: size distribution interpretation. J. Air Pollut. Control Assoc. **25**(5), 529–534 (1975)
28. U. Pöschl, Atmospheric aerosols: composition, transformation, climate and health effects. Angew. Chem. Int. Ed. **44**(46), 7520–7540 (2005)
29. F.-M. Bréon, D. Tanré, S. Generoso, Aerosol effect on cloud droplet size monitored from satellite. Science **295**(5556), 834–838 (2002)
30. S. Twomey, The influence of pollution on the shortwave albedo of clouds. J. Atmos. Sci. **34**(7), 1149–1152 (1977)
31. G. Feingold, W.L. Eberhard, D.E. Veron, M. Previdi, First measurements of the twomey indirect effect using ground-based remote sensors. Geophys. Res. Lett. **30**(6), 1287 (2003)
32. M. Ligda, Meteorological observations with a pulsed laser radar, in *Proceedings 1st Conference on Laser Technology* (U.S. Office of Naval Research), pp. 63–72 (1963)
33. G. Fiocco, L.D. Smullin, Detection of scattering layers in the upper atmosphere (60–140 km) by optical radar. Nature **199**(4900), 1275–1276 (1963)
34. G. Fiocco, G. Grams, Observations of the aerosol layer at 20km by optical radar. J. Atmos. Sci. **21**, 323–324 (1964)
35. T.H. Maiman, Stimulated optical radiation in ruby. Nature **187**(4736), 493–494 (1960)
36. J. Reichardt, S. Reichardt, Determination of cloud effective particle size from the multiple-scattering effect on lidar integration-method temperature measurements. Appl. Opt. **45**(12), 2796–2804 (2006)
37. T. Berkoff, E. Welton, J. Campbell, S. Valencia, J. Spinhirne, S.-C. Tsay, B. Holben, Observations of aerosols using the Micro-Pulse Lidar NETwork (MPLNET), in *Proceedings of IEEE International Geoscience and Remote Sensing Symposium, 2004*, vol. 3, pp. 2208–2211 (2004)
38. O. Uchino, M. Tokunaga, M. Maeda, Y. Miyazoe, Differential-absorption-lidar measurement of tropospheric ozone with excimer-Raman hybrid laser. Opt. Lett. **8**(7), 347–349 (1983)
39. G. Vaughan, D.P. Wareing, S.J. Pepler, L. Thomas, V. Mitev, Atmospheric temperature measurements made by rotational Raman scattering. Appl. Opt. **32**(15), 2758–2764 (1993)
40. T. Mikkelsen, J. Mann, M. Courtney, M. Sjholm, Windscanner: 3-D wind and turbulence measurements from three steerable Doppler lidars. IOP Conf. Ser. Earth Environ. Sci. **1**(1), 012018 (2008)
41. D. Roberts, G. Gimmestad, Optimizing lidar dynamic range by engineering the crossover region, in *Proceedings of 21st International Laser Radar Conference* (2002)
42. E. Gregorio, F. Rocadenbosch, J. Tiana-Alsina, A. Comern, R. Sanz, J.R. Rosell-Polo, Parameter design of a biaxial lidar ceilometer. J. Appl. Remote Sens. **6**(1), 063546-1–063 546–19 (2012)
43. J. Spinhirne, Micro pulse lidar. IEEE Trans. Geosci. Remote Sens. **31**(1), 48–55 (1993)
44. G. Mie, Beiträge zur optik trüber medien, speziell kolloidaler metallösungen. Ann. Phys. **330**(3), 377–445 (1908)
45. E. Clothiaux, G. Mace, T. Ackerman, T. Kane, An automated algorithm for detection of hydrometeor returns in micropulse lidar data. J. Atmos. Oceanic Technol. **15**(8), 1035–1042 (1998)
46. A. Lavrov, A.B. Utkin, R. Vilar, A. Fernandes, Evaluation of smoke dispersion from forest fire plumes using lidar experiments and modelling. Int. J. Therm. Sci. **45**(9), 848–859 (2006)

47. J.L. Gaumet, J.C. Heinrich, M. Cluzeau, P. Pierrard, J. Prieur, Cloud-base height measurements with a single-pulse erbium-glass laser ceilometer. J. Atmos. Oceanic Technol. **15**(1), 37–45 (1998)
48. X. Xiong, S. Feng, L. Jiang, Z. Zhuang, Research on lidar visibility measurement system for airport, in *2011 International Conference on Electric Information and Control Engineering (ICEICE)*, pp. 3514–3517, April 2011
49. D.L. Hutt, L.R. Bissonnette, L. Durand, Multiple field of view lidar returns from atmospheric aerosols. Appl. Opt. **33**(12), 2338–2348 (1994)
50. C. Münkel, Rain-snow discrimination with a biaxial lidar ceilometer. 12th international workshop on lidar multiple scattering experiments. SPIE the international society for optical engineering, Vol. 5059, pp. 160–170 (2003)
51. D. Jenkins, D. Wareing, L. Thomas, G. Vaughan, Upper stratospheric and mesospheric temperatures derived from lidar observations at Aberystwyth. J. Atmos. Terr. Phys. **49**(3), 287–298 (1987)
52. M. Nicolet, M. Schnaiter, O. Stetzer, Circular depolarization ratios of single water droplets and finite ice circular cylinders: a modeling study. Atmos. Chem. Phys. **12**(9), 4207–4214 (2012)
53. K. Sassen, The polarization lidar technique for cloud research: a review and current assessment. Bull. Am. Meteorol. Soc. **72**, 1848–1866 (1991)
54. G.G. Gimmestad, Reexamination of depolarization in lidar measurements. Appl. Opt. **47**(21), 3795–3802 (2008)
55. T. Sakai, T. Nagai, M. Nakazato, Y. Mano, T. Matsumura, Ice clouds and Asian dust studied with lidar measurements of particle extinction-to-backscatter ratio, particle depolarization, and water-vapor mixing ratio over Tsukuba. Appl. Opt. **42**(36), 7103–7116 (2003)
56. N. Segre, J.P. Truscott, Erbium laser ceilometer, US Patent 3 963 347, 6 15 (1976)
57. W.L. Eberhard, Cloud signals from lidar and rotating beam ceilometer compared with pilot ceiling. J. Atmos. Oceanic Technol. **3**(3), 499–512 (1986)
58. C. Flynn, ARM technical report TR-040: belfort laser ceilometer handbook. US Department of Energy Atmospheric Radiation Measurement (ARM), Technical Report (2004)
59. Eliasson Engineering, AB, CBME80 data sheet (2012), http://www.eliasson.com/products/cbme80.shtml
60. All Weather, Inc. (AWI) 8339 and 8440 laser cloud height brochures (2012), http://www.allweatherinc.com/meteorological-sensors/cloud-height-ceilometers/ (August 2014)
61. C. Münkel, J. Räsänen, New optical concept for commercial lidar ceilometers scanning the boundary layer, vol. 5571, no. 1, ed. by A. Comeron, M.R. Carleer, R.H. Picard, N.I. Sifakis, SPIE, pp. 364–374 (2004)
62. Vaisala, Oyj, CL31 datasheet (2012), http://www.vaisala.com/en/products/ceilometers/Pages/cl31.aspx
63. G. Martucci, C. Milroy, C.D. ODowd, Detection of cloud-base height using Jenoptik CHM15k and Vaisala CL31 ceilometers. J. Atmos. Oceanic Technol. **27**(2), 305–318 (2010)
64. C. Münkel, N. Eresmaa, J. Räsänen, A. Karppinen, Retrieval of mixing height and dust concentration with lidar ceilometer. Bound.-Layer Meteorol. **124**, 117–128 (2007). doi:10.1007/s10546-006-9103-3
65. G. Tsaknakis, A. Papayannis, P. Kokkalis, V. Amiridis, H.D. Kambezidis, R.E. Mamouri, G. Georgoussis, G. Avdikos, Inter-comparison of lidar and ceilometer retrievals for aerosol and planetary boundary layer profiling over Athens, Greece. Atmos. Meas. Tech. **4**, 1261–1273 (2011)
66. A. Kaerkkaeinen, A. Piironen, T. Kaehkoenen, and J. Loennqvist, Characteristics and performance of Vaisala's new CT75K lidar ceilometer, in *Society of Photo-Optical Instrumentation Engineers (SPIE) Conference Series*, ed. by J.-P. Wolf. Society of Photo-Optical Instrumentation Engineers (SPIE) Conference Series, vol. 3104, pp. 12–17, May 1997
67. O. Krasnov, H. Russchenberg, A synergetic radar-lidar technique for the LWC retrieval in water clouds, in *7th International Symposium on Tropospheric Profiling: Needs and Technologies*, Boulder, CO, USA, 11–16 June 2006

68. E. O'Connor, A. Illingworth, R. Hogan, A technique for autocalibration of cloud lidar. J. Atmos. Oceanic Technol. **21**, 777–786 (2004)
69. Vaisala, Oyj, CL51 datasheet (2012), http://www.vaisala.com/en/products/ceilometers/Pages/CL51.aspx
70. D. Engelbart, J. Reichardt, G. Teschke, Intercomparison of methods for determining mixing height using a new network of single-photon-counting high-sensitivity ceilometers in Germany, in *Proceedings of WMO meeting, December, 2008, St. Petersburg, Russian Federation* (2008)
71. M. Wiegner, A. Geiß, Aerosol profiling with the Jenoptik ceilometer CHM15kx. Atmos. Measur. Tech. **5**(8), 1953–1964 (2012)
72. Leosphere, R-MAN 510 Super Ceilometer Datasheet (2012), http://leosphere.com/page.php?rubrique=182
73. Degreane Horizon, Laser ceilometer datasheet, model ALC 30 (2012), http://www.degreane-horizon.com/GB/meteorology/autonomous-measurements/cloud-height-measurement.html (August 2014)
74. MTECH Systems, 8200-CHS datasheet (2012), http://www.mtechsystems.com/data-sheets/8200-CHS-Ceilometer-Sensor.pdf
75. Tempestini, Model TMP-09-03 specification (2012), http://www.tempestinisystems.it (August 2014)
76. E. Gregorio, F. Rocadenbosch, A. Comeron, 905-nm biaxial lidar ceilometer prototype, in *Proceedings of Remote Sensing of Clouds and the Atmosphere XI*, ed. by J.R. Slusser, K. Schafer, A. Comeron, vol. 6362, no. 1. SPIE, p. 63621L (2006)
77. E. Gregorio, F. Rocadenbosch, A. Comeron, Design methodology of a ceilometer lidar prototype, in *Proceedings of IEEE International Geoscience and Remote Sensing Symposium, 2007*, pp. 3162–3165 (2007)
78. M. Wiegner, S. Emeis, V. Freudenthaler, B. Heese, W. Junkermann, C. Mnkel, K. Schfer, M. Seefeldner, S. Vogt, Mixing layer height over Munich, Germany: variability and comparisons of different methodologies. J. Geophys. Res. **111**(D13), D13 201 (2006)
79. E.W. Weisstein, Normal Distribution (2012). From MathWorld-A Wolfram Web Resource, http://mathworld.wolfram.com/NormalDistribution.html
80. I.M. Brooks, Finding boundary layer top: Application of a wavelet covariance transform to lidar backscatter profiles. J. Atmos. Oceanic Technol. **20**(8), 1092–1105 (2003)
81. R.J. Teschke, D. Engelbart, Wavelet algorithm for the estimation of mixing layer height with ceilometers, in *Reviewed and revised papers presented at the 24th International Laser Radar Conference*, p. 313316 (2008)
82. M. de Haij, W. Wauben, H. Klein Baltink, A. Apituley, Determination of the mixing layer height by a ceilometer, in *Proceedings of the 8th International Symposium on Tropospheric Profiling, ISTP-2009*, ed. by W.M.A. Apituley, H.W.J. Russchenberg (2009)
83. M. Haeffelin, F. Angelini, Y. Morille, G. Martucci, S. Frey, G. Gobbi, S. Lolli, C. O'Dowd, L. Sauvage, I. Xueref-Rémy, B. Wastine, D. Feist, Evaluation of mixing-height retrievals from automatic profiling lidars and ceilometers in view of future integrated networks in Europe. Bound.-Layer Meteorol. **143**, 49–75 (2012)
84. F. Di Giuseppe, A. Riccio, L. Caporaso, G. Bonaf, G.P. Gobbi, F. Angelini, Automatic detection of atmospheric boundary layer height using ceilometer backscatter data assisted by a boundary layer model. Q. J. R. Meteorol. Soc. **138**(664), 649–663 (2012)
85. C. Münkel, S. Emeis, W. Müller, and K. Schäfer, Aerosol concentration measurements with a lidar ceilometer: results of a one year measuring campaign, in *Remote Sensing of Clouds and the Atmosphere VIII*, vol. 5235, ed. by K.P. Schäfer, A. Comerón, M.C. adn Richard H. Picard (2004)
86. B. Heese, H. Flentje, D. Althausen, A. Ansmann, S. Frey, Ceilometer lidar comparison: backscatter coefficient retrieval and signal-to-noise ratio determination. Atmos. Meas. Tech. **3**(6), 1763–1770 (2010)
87. I. Stachlewska, M. Piadłowski, S. Migacz, A. Szkop, A. Zielińska, P. Swaczyna, Ceilometer observations of the boundary layer over Warsaw, Poland. Acta Geophys. **60**, 1386–1412 (2012). doi:10.2478/s11600-012-0054-4

Chapter 2
Theory of Lidar

2.1 Introduction

This chapter introduces the subject of light scattering by molecules and water droplets in the atmosphere and gives an overview of the theory essential for interpreting elastic lidar returns from clouds and aerosols. First, parameters describing droplet size distributions in clouds are explained. Following this, scattering, extinction, and the relationship between the two are considered. Then the elastic backscatter lidar equation is introduced and the derivation of a standard signal inversion method, commonly referred to as the Klett inversion, is given, along with some discussion of its strengths and shortcomings. Finally, an expression for attenuated backscatter, the range-corrected calibrated signal, is given for use in comparison of measurements from different instruments.

2.2 Composition of Liquid Clouds

Because the nature of the particles that constitute a cloud determines the way light will interact with it, it is important to consider cloud composition when probing with lidar. Cloud properties have been studied extensively through active and passive remote sensing techniques, in situ measurements, and modelling [1]. This knowledge, in the form of cloud droplet size distributions and particle concentrations, for example, can be used to model how light is scattered by a cloud and predict return signals that might be detected from it. This is important for the current work particularly because water clouds with well understood droplet properties can be used for ceilometer attenuated backscatter calibration as explained later in this chapter and applied in Chap. 5.

A typical approach to modelling cloud droplet sizes is to consider spherical droplets with the gamma type distribution function, $f(a)$, described by Deirmendjian such that [2]

J.D. Vande Hey, *A Novel Lidar Ceilometer*, Springer Theses,
DOI 10.1007/978-3-319-12613-5_2

$$f(a) = \frac{\mu^{\mu+1} a^{\mu} e^{-\mu \frac{a}{a_0}}}{\Gamma(\mu+1) a_0^{\mu+1}} \tag{2.1}$$

where a is a random variable representing droplet radius, a_0 is the mode of the distribution, and μ describes the width of the distribution and can be expressed as

$$\mu = \frac{1}{C_\sigma^2} - 1, \tag{2.2}$$

in which C_σ is the coefficient of variance. Note that the function $\Gamma(x)$ is the gamma function [3]

$$\Gamma(x) = \int_0^\infty e^{-t} t^{x-1} dt, \tag{2.3}$$

which for integer values of x can be calculated as a factorial such that

$$\Gamma(x+1) = x! \tag{2.4}$$

Often the parameter used to describe the droplet size distribution is the effective radius, a_{eff}, where [4]

$$a_{\text{eff}} = (1 + \frac{3}{\mu}) a_0. \tag{2.5}$$

For water clouds the mode, a_0, ranges from 4 to 20 μm [5], and μ ranges from 2 to 8, which produces an effective radius range of 5–50 μm [4]. According to Han et al., however, a_{eff} typically ranges from 5 to 15 μm [6].

The C1 distribution of Deirmendjian [2], commonly used to model the droplet size distribution of cumulus clouds in the literature, uses the values $a_0 = 4$ μm and $\mu = 6$ in Eq. 2.1. This yields an effective radius of $a_{\text{eff}} = 6$ μm. Fomin and Mazin [7] caution that when relating model to measurement it is important to consider that the width of the droplet size distribution typically increases with the volume of the sampling region. They suggest that a value of $\mu = 6$ only applies to small spatial averaging regions.

Figure 2.1 shows an example of close agreement between measured and modelled distributions. Here the results of an in situ measurement of droplet size distribution using a forward scattering spectrometer probe 500 m above the base of a continental cumulus cloud given by [8] were extracted from the original data and plotted alongside the modelled C1 distribution.

A comprehensive discussion of the formation and makeup of liquid water, mixed phase, and ice clouds is beyond the scope of this work. Understanding these properties and processes is an important area of research, however, both in terms of simulation and inversion of lidar returns, and in terms of improving understanding of radiative

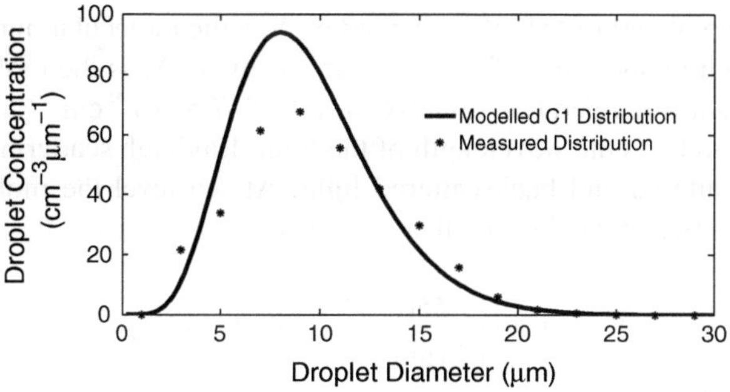

Fig. 2.1 Droplet diameter distribution measured 500 m above cloud base in continental cumulus in Northern Colorado, USA, with a measured number density $N = 780\,\mathrm{cm}^{-3}$, as reported by Knollenberg [8] compared with the C1 distribution at the same number density

processes in clouds. These subjects are considered in detail in works by Pruppacher and Klett [1], Hobbs and Deepak [9], and Lynch et al. [10].

2.3 Elastic Scattering and Transmission of Light in the Atmosphere

Elastic interaction of light with scatterers is typically described by one of two processes, depending upon the size of the scatterers. Rayleigh scattering is used to describe scattering in situations where the scatterer is very small compared to the wavelength of the light. Mie scattering is typically used to describe interactions of light with particles whose sizes are similar to or somewhat larger than the wavelength of the light. Note that single-scattering, in which a photon interacts with one scatterer only before being detected, is typically the dominant process measured by lidar. However, multiple scattering, in which a photon interacts with more than one scatterer before being detected, must often be considered, particularly in systems with large fields of view.

2.3.1 Rayleigh Scattering

Scattering by gas molecules in the atmosphere can be described by Rayleigh scattering. As expressed by Kovalev and Eichinger [11], the wavelength-dependent molecular volume backscatter coefficient β_m can be determined such that

$$\beta_m = \frac{8\pi^3(m^2-1)N}{3{N_s}^2\lambda^4}, \tag{2.6}$$

where m is the real part of the refractive index, N is the molecular number density at the pressure and temperature of the scattering volume, N_s is the molecular number density at standard temperature and pressure ($2.547 \times 10^{19}\,\mathrm{cm}^{-3}$ at 288.15 K and 101.325 kPa), and λ is the wavelength of the light. Rayleigh scattering is symmetric for forward scattered and backscattered light. At sea level the molecular volume backscatter coefficient can be calculated as [11]

$$\beta_m = 1.39\left[\frac{550}{\lambda(\mathrm{nm})}\right]^4 \times 10^{-8}\,\mathrm{cm}^{-1}\mathrm{sr}^{-1}. \tag{2.7}$$

The most significant factor in these expressions is the λ^{-4} wavelength dependence of the scattering intensity which means, for example, that ultraviolet light at 355 nm is scattered 81 times more strongly by the molecular atmosphere than near-infrared light at 1,064 nm. Note that while scattering by very small particles is in fact Rayleigh scattering, in lidar research the term Rayleigh scattering usually refers only to scattering from the molecular atmosphere [12].

2.3.2 Mie Scattering

When the scatterer size is similar to the wavelength of incident light, a theory developed by Gustav Mie in 1908 [13] can be used to calculate the scattering phase function for spherical, optically conducting particles. In this case the scattering is given by an infinite series expansion.

Some example scattering phase functions calculated by a Mie-based method [14] at a wavelength of 1 μm for spheres of various radii with refractive index 1.5 are shown in Figs. 2.2 and 2.3. As shown in Fig. 2.2, when the radius is 0.1 μm, ten times smaller than the wavelength, the forward and backward scattering distributions are similar to each other. Note that when the radius is decreased to 0.01 μm, the forward and backward scattering distributions become essentially symmetric and approach the Rayleigh scattering solution.

As the particle radius increases with respect to the wavelength, the distribution becomes more pointed in the forward direction. Figure 2.3 shows the scattering distribution for radius equal to wavelength. A close-up view around the origin shows scattering lobes at various angles, the position and intensity of which relate to interference patterns of light propagating around and through the sphere; these will vary from the theory for non-spherical shapes and imperfect optical conductor materials [11]. As the size of the sphere increases further, Mie theory gives larger and larger forward scattering lobes until the sphere is significantly larger than the wavelength of the light, at which time geometric optics [15] can be used to describe the light path. Note that Deirmendjian [2] used Mie theory to calculate scattering phase functions for C1 and other droplet size distributions, thus providing a reference that has been used, for example, to calculate multiple scattering effects in clouds [16].

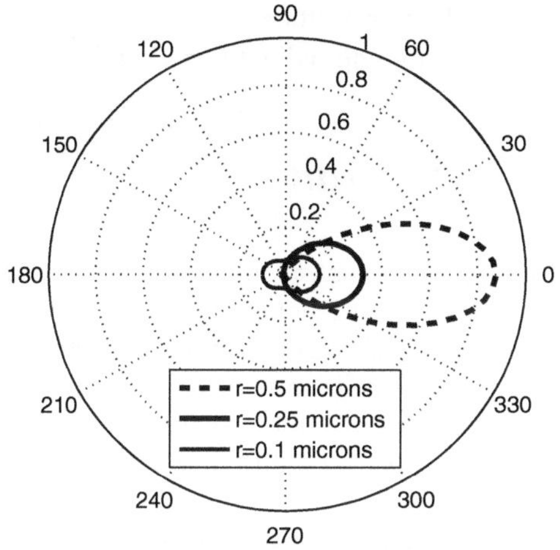

Fig. 2.2 Relative scattering intensity of 1 μm wavelength light by spherical particles of refractive index 1.5 as a function of angle (in degrees) for three different radii considering light entering from the *left*

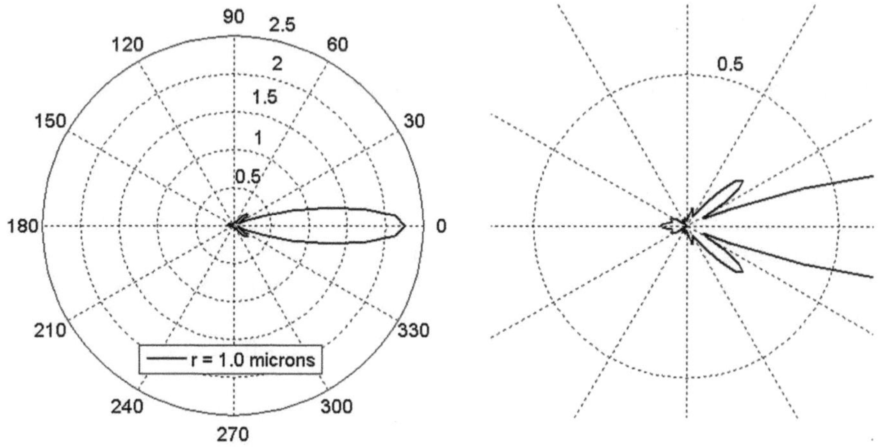

Fig. 2.3 Relative scattering intensity of 1 μm wavelength light by spherical particles of refractive index 1.5 as a function of angle (in degrees) for particles of 1 μm radius considering light entering from the *left*. A close-up view around the origin is shown on the *right*

2.3.3 Transmission and Extinction

The transmission and extinction of light travelling through the atmosphere are key considerations for lidar. They are explained here following the approach of Kovalev and Eichinger [11].

For a given wavelength, the transmittance $T(H)$ of a layer of thickness H can be expressed as the ratio of the outgoing radiant flux F to the incoming radiant flux F_0 such that

$$T(H) = \frac{F}{F_0}. \tag{2.8}$$

$T(H)$ ranges from 0 for a fully-attenuating medium to 1 for a medium through which all of the light passes without experiencing any scattering or absorption. In order to account for range-variable transmission through a heterogeneous medium, the extinction coefficient function $\alpha(r)$ is introduced to describe, for each differential range element dr, the probability of photon scattering or absorption per unit path length. The change in radiant flux over a differential element can be considered as a function of $\alpha(r)$ such that

$$dF(r) = -\alpha(r)F(r)dr. \tag{2.9}$$

From this expression the Beer-Lambert-Bougert law, which relates outgoing to incoming radiant flux, can be derived such that

$$F = F_0 e^{-\int_0^H \alpha(r)dr}, \tag{2.10}$$

and by substituting Eq. 2.8 into Eq. 2.10 the transmittance can then be expressed as

$$T(H) = e^{-\int_0^H \alpha(r)dr}. \tag{2.11}$$

Here the integral in the exponent $\int_0^H \alpha(r)dr$ is the summed extinction along the path and is therefore used to express the optical depth τ_α.

If inelastic scattering is sufficiently small to be disregarded as is usually the case for elastic lidar, the extinction coefficient can be expressed as the sum of the total elastic scattering coefficient, $\beta_T(r)$, and the absorption coefficient, $\alpha_A(r)$, such that

$$\alpha(r) = \beta_T(r) + \alpha_A(r). \tag{2.12}$$

For a cloud lidar, liquid water droplets are the primary particles of interest. The complex refractive index of water (the real part of which is approximately 1.33), which can be used to determine its scattering and absorption properties for a given radius, has been measured over a wide range of wavelengths by various authors, for example, by Hale and Querry [17]. However, total absorption and scattering vary with size and concentration of scatterers, which may not be known, and soluble aerosols dissolved in water may add complexity to light-particle interactions by increasing the imaginary part of the refractive index, i.e. absorption. In addition, the presence of mixed particle types in a scattering volume may introduce further complexity. This means it can be very difficult to separate the constituents of the extinction coefficient. Nonetheless, by applying some assumptions about the atmosphere and/or or by including information from external measurements, it is possible to draw some conclusions about the relationship between total extinction and backscatter, a relationship that is of key importance to lidar measurements.

2.3.4 The Lidar Ratio

Interpreting the physical meaning of a measured lidar signal is an inverse problem. Inversion techniques must therefore be applied in order to determine optical properties of atmospheric constituents from which return signals are collected. From an elastic lidar measurement of range-resolved power, it is not possible to distinguish with certainty the contributions of the two variables, extinction and backscatter, to the signal profile because the relationship between the range-dependent backscatter coefficient $\beta(r)$ and the range-dependent extinction coefficient $\alpha(r)$ varies depending upon the content of the measurement volume at each range r. Consider, for example, that a return from a thin, diffuse, weakly scattering cloud layer with clear air between it and the lidar instrument could look very similar to a return from thin, dense, strongly scattering cloud layer with a strongly absorbing gas layer between it and the lidar. In order to account for these differences, the range-dependent backscatter to extinction ratio $\Pi_p(r)$ can be expressed [11]

$$\Pi_p(r) = \beta(r)/\alpha(r). \tag{2.13}$$

This expression can be useful for applying assumptions to the lidar ratio over the detection range if, for example, $\Pi_p(r)$ is assumed to be constant or a linear function, or if it can be calculated using a model or measured by some method.

Klett explained that the relationship between backscatter and extinction can be also be approximately expressed in the form [18]

$$\beta(r) = B_0\alpha^k(r), \tag{2.14}$$

where B_0 and k are assumed to be constants. This power law expression was used in differential form in Klett's original derivation of the backward inversion method. He noted that k is wavelength-dependent and also influenced by aerosol properties in the measurement volume and explained that it is typically in the range of $0.67 \leq k < 1.0$.

The lidar ratio is a fundamental unknown for most elastic lidar measurements. The quality of elastic lidar inversion often depends on the accuracy of the assumptions made regarding the lidar ratio.

2.4 Elastic Lidar System Constant

A number of parameters of a lidar system that affect the measured level of backscattered light are typically factored into a system constant, K_s, expressed by Wandinger [12] such that

$$K_s = P_0 \frac{c\tau}{2} A_0 \eta. \tag{2.15}$$

Here P_0 is the average laser power output during a pulse, c is the speed of light, τ is the laser pulse duration, A_0 is the area of the receiver objective as shown in Fig. 2.4, and η is the total efficiency of the instrument's optical path multiplied by the

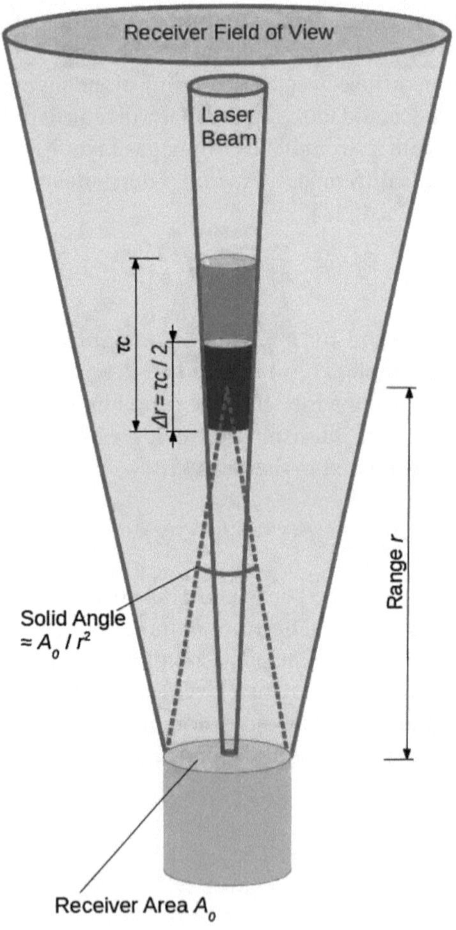

Fig. 2.4 Lidar geometry

detection efficiency. The value of P_0 determines the peak intensity of the backscatter cross section. A_0 determines the solid angle of the backscatter cross section that is subtended from a given range as shown in Fig. 2.4. The factor $\frac{c\tau}{2}$, also illustrated in Fig. 2.4, determines the range resolution of a pulsed system by establishing the time (distance) required for backscattered light from the beginning of the pulse to meet the forward-propagating light from the end of the pulse. Typically τ and A_0 are well characterised, P_0 may be known but is often prone to variability, and η can usually only be estimated to varying degrees of accuracy depending upon the complexity of the system and understanding of the efficiency of each component. Because of this uncertainty, it is often desirable to find a way to process lidar using a method that allows dependence on specific knowledge of the system constant to be cancelled out.

2.5 The Single-Scattering Elastic Lidar Equation

In the case where molecular returns are negligible compared to aerosol returns, which is the realm most relevant to ceilometer measurements, the range-dependent backscatter and extinction coefficients can be considered as functions only of the aerosol and water droplet returns. Under the assumption of single-scattering, and if a discrete laser wavelength is used, the lidar equation in this case can be expressed in the form [12]

$$P(r) = K_s \beta(r) \frac{O(r)}{r^2} e^{-2 \int_0^r \alpha(r')dr'}, \tag{2.16}$$

where $P(r)$ is the power detected from range r, K_s is the system constant given in Eq. 2.15, $\beta(r)$ is the scattering coefficient from the scattering volume at range r, $O(r)$ is the overlap function which describes what fraction of the laser beam cross section at a given range will be imaged onto the detector (this function reaches a value of unity at a full overlap distance r_0), $\frac{1}{r^2}$ is the range dependence factor that accounts for the decrease in solid angle subtended with the square of the range, and the remaining factor, $e^{-2 \int_0^r \alpha(r')dr'}$, based on the range dependent extinction coefficient $\alpha(r)$, is the integrated two-way extinction of the signal as it propagates from the instrument to the scattering volume at range r and back.

2.6 Elastic Lidar Inversion

A number of approaches for the inversion of elastic lidar signals have been described in the literature. Each of these methods applies a different set of assumptions in order to achieve inversion. While there are a variety of alternatives to, variations on, and combinations of these techniques (for example, [19, 20]) as well as detailed error analyses (for example, [21, 22]), only the fundamental methods are discussed here.

2.6.1 Slope Method

The first lidar inversion method described in the literature was the slope method discussed in 1966 by Collis [23]. In homogeneous atmospheric conditions, the extinction and backscatter coefficients can be assumed to be constants. In this case, the natural logarithm of the range-corrected lidar return is linear and from its slope the extinction coefficient can be derived. Kovalev and Eichinger note that in order to satisfy the homogeneity requirement the atmosphere need not be purely homogeneous but rather that local inhomogeneities do not significantly alter the linear fit across the region of interest [11]. They explain that for homogeneous atmospheres this method is often the best way to extract mean aerosol extinction, particularly if the aerosol and molecular returns are of similar amplitude. Kovalev and Eichinger caution, however, that if returns from aerosol-free atmospheres are being processed by this method, care must be taken to fully account for any background noise on the signal as that can greatly affect the slope that is calculated. They also emphasise the importance of either disregarding the region of incomplete overlap of the transmitter and receiver or carefully compensating for it.

2.6.2 Close Boundary Solution

A close boundary solution applies an assumed or measured value of the extinction coefficient at the start of the measurement range and inverts the signal in the forward direction. This method was first applied to lidar in 1967 by Barrett and Ben-Dov [24]. It can be applied successfully in clear atmospheric conditions, but in turbid conditions it quickly becomes unstable due to its mathematical formulation; its performance can be improved somewhat, however, by placing constraints that limit the possible solutions to positive values of extinction without extremely large "runaway" values [25].

2.6.3 Optical Depth Solution

Another approach to inversion, first introduced in 1988 by Weinman [26], is the optical depth solution. If the total optical depth of a lidar measurement range can be estimated, the transmission term in the lidar equation can be determined; this then acts as a constraint for the inversion. An important calibration method described in Sect. 2.8 is related to this approach. In order to perform an inversion based on the optical depth solution for combined molecular-aerosol atmospheres, three inputs typically used are the sun photometer-derived aerosol optical depth, the profile of molecular extinction (this may be disregarded if aerosol extinction is much greater than molecular extinction at the laser wavelength), and an aerosol lidar ratio assumption relating backscatter to extinction [27].

2.6.4 Far Boundary Solution

The most widely used method for inverting elastic lidar returns is the backward inversion method. This method, though also developed in similar form independently by Kaul in 1977 [28] and Zuev in 1978 [29]–those studies were not accessible in Western countries at the time–is typically credited to Klett who introduced it in 1981 [18]. In this method the extinction coefficient at the far boundary is assumed and the signal is inverted backward toward the instrument. This approach provides a stable result provided there are considerable aerosol or cloud returns present and is therefore the method typically applied in the inversion of ceilometer returns. The contemporary version of Klett's approach was reformulated in 1982 by Fernald [30], and a method for improving the lidar ratio assumption and a smoothing process at the boundary point were described in 1984 by Sasano and Nakane [31]. It is therefore sometimes referred to as the Klett-Fernald-Sasano inversion, however, in this thesis it is simply referred to as the Klett inversion.

The Klett inversion requires an input value of the extinction coefficient at the far boundary of the lidar range. This boundary value can be measured or assumed. Since information from in situ measurement of the extinction coefficient at the far range of the instrument is not usually available for vertical lidar profiling, boundary extinction is typically assumed from some knowledge of the current atmospheric conditions.

2.6.5 Derivation of the Far Boundary Solution

Under the assumption of a single-component atmosphere, in which aerosol returns dominate molecular returns (reasonable for ceilometers at 905 nm), the far-boundary solution can be derived in a straightforward manner following the approach of Kovalev and Eichinger [11]. The lidar equation can first be rewritten somewhat by removing the overlap dependence. Overlap and its correction are discussed in detail in Chaps. 3 and 4, but here only ranges beyond the full overlap height r_0 are considered such that $O(r) = 1$. Equation 2.16 therefore becomes

$$P(r) = K_s {T_{r_0}}^2 \frac{\beta(r)}{r^2} e^{-2\int_{r_0}^{r} \alpha(r')dr'}, \tag{2.17}$$

where ${T_{r_0}}^2$ is a constant accounting for the unknown transmittance from $r = 0$ to $r = r_0$.

If the lidar ratio is introduced as in Eq. 2.13, the backscatter coefficient $\beta(r)$ can be expressed in terms of extinction $\alpha(r)$ and lidar ratio function $\Pi_p(r)$ allowing Eq. 2.17 to be rewritten as

$$P(r) = K_s {T_{r_0}}^2 \frac{\Pi_p(r)\alpha(r)}{r^2} e^{-2\int_{r_0}^{r} \alpha(r')dr'}. \tag{2.18}$$

Assuming that the particles along the measurement range are similar to each other, the backscatter to extinction ratio can be expressed by the constant Π_ρ such that

$$\Pi_p(r) = \Pi_p. \tag{2.19}$$

Kovalev and Eichinger note that if the variation among scatterers is relatively small, this assumption is reasonable, but that if precise determination of extinction is required, it can be problematic. If the assumption is applied, the unknown parameters of the measurement can now be expressed as a single constant such that

$$K_L = K_s {T_{r0}}^2 \Pi_p, \tag{2.20}$$

and Eq. 2.18 can be written as

$$P(r) = K_L \frac{\alpha(r)}{r^2} e^{-2\int_{r0}^{r} \alpha(r')dr'}. \tag{2.21}$$

If $Z(r)$, the range-corrected signal, is considered, where $Z(r) = P(r)r^2$, then

$$Z(r) = K_L \alpha(r) e^{-2\int_{r0}^{r} \alpha(r')dr'}. \tag{2.22}$$

If the extinction coefficient $\alpha(r_b)$ at the far boundary r_b is known or can be estimated, the constant K_L can be written

$$K_L = \frac{Z(r_b)}{\alpha(r_b) e^{-2\int_{r0}^{rb} \alpha(r')dr'}}. \tag{2.23}$$

Substituting this into Eq. 2.22, the range-corrected signal becomes

$$Z(r) = \frac{Z(r_b)\alpha(r) e^{-2\int_{r0}^{r} \alpha(r')dr'}}{\alpha(r_b) e^{-2\int_{r0}^{rb} \alpha(r')dr'}}. \tag{2.24}$$

If this is expressed as the ratio of range corrected signal $Z(r_b)$ at the boundary r_b to the extinction coefficient $\alpha(r_b)$ at the boundary,

$$\frac{Z(r_b)}{\alpha(r_b)} = \frac{Z(r)}{\alpha(r)} \frac{e^{-2\int_{r0}^{rb} \alpha(r')dr'}}{e^{-2\int_{r0}^{r} \alpha(r')dr'}}, \tag{2.25}$$

which can be rewritten as

$$\frac{Z(r_b)}{\alpha(r_b)} = \frac{Z(r)}{\alpha(r)} e^{-2\int_{r}^{rb} \alpha(r')dr'}. \tag{2.26}$$

By solving this equation for $Z(r)$ and integrating both sides over the interval r to r_b, an expression for the transmission term, $e^{-2\int_r^{r_b}\alpha(r')dr'}$, can be derived in the form

$$e^{-2\int_r^{r_b}\alpha(r')dr'} = 1 + \frac{\alpha(r_b)}{Z(r_b)}\left[2\int_r^{r_b} Z(r')dr'\right]. \tag{2.27}$$

If this is substituted into Eq. 2.26 and $\alpha(r)$ is solved for, the Klett solution is arrived at such that

$$\alpha(r) = \frac{Z(r)}{\frac{Z(r_b)}{\alpha(r_b)} + 2\int_r^{r_b} Z(r')dr'}. \tag{2.28}$$

In this way the range-dependent extinction coefficient is expressed in terms of only the range-corrected signal and the boundary value of the extinction coefficient.

This solution is useful because even though it relies on an assumed boundary condition at the far boundary, it is stable in the near-range due to the fact that the factor $2\int_r^{r_b} Z(r')dr'$ increases with decreasing r. In addition, as ranges closer and closer to the instrument are considered, errors due to incorrect assumption of the far boundary condition reduce significantly due to the reduced influence of the factor $\frac{Z(r_b)}{\alpha(r_b)}$, particularly in turbid atmospheres [18]. This far boundary solution is applied in Chap. 5 in order to invert prototype lidar returns.

2.6.6 Boundary Condition Selection for Klett Inversion

There are two factors to consider when making the far-range boundary assignment. The first of these is the signal level. Usually the maximum possible measurement range is desirable. In this case, the boundary condition is assigned at the farthest point in the signal that is above a certain threshold. Two examples of this from the literature are 2.3 % of the maximum digitised signal amplitude [32], assuming the dynamic range of the receiver electronics is appropriately matched to the dynamic range of the signals, or, in another work, a signal to noise ratio of 5–10 dB [33]. Choosing a boundary point too close to the noise level is likely to reduce the accuracy of the inversion. A signal to noise ratio method for locating the boundary range is applied in Chap. 5.

The second factor to consider is the boundary value of the extinction coefficient. It is possible to estimate the boundary value of the extinction coefficient directly from the signal by considering the slope of the logarithmic range-corrected signal as explained by Klett [18], such that

$$\alpha_b \approx \frac{\ln\left[P(r_s){r_s}^2\right] - \ln\left[P(r_b){r_b}^2\right]}{2(r_b - r_s)}. \tag{2.29}$$

This is calculated over a region starting at a selected range r_s and reaching to the maximum range r_b at the boundary. This method works best over a homogeneous region with significant returns.

If no clearly homogeneous region is present, a default value can be used depending upon measurement conditions. While a clear air value such as $10^{-4}\,\mathrm{m}^{-1}$ can be applied as in [34], for cloud detection applications it may be better to select a cloud value such as $0.02\,10^{-4}\,\mathrm{m}^{-1}$ as in [32], particularly if it is unlikely that clear air can be detected at far ranges or beyond a cloud layer due to attenuation and lack of sensitivity. Kovalev and Eichinger state that if the effects of multiple scattering are small, it is straightforward to assign a boundary value $\alpha(r_b)$ within a cloud. This approach is generally applicable to ceilometers, both because of their limited sensitivity to molecular returns and because their primary function is cloud detection, and is applied in Chap. 5.

2.6.7 *Stability of the Klett Inversion*

A few examples of the behaviour of the Klett inversion under various conditions illustrate its value. First, Fig. 2.5 shows the influence on the inversion of overestimating or underestimating the boundary value of the extinction coefficient by 50 %. In both cases the inversion converges on the correct profile quite quickly as the range is considered backwards from the boundary r_b toward $r = 0$. Second, Fig. 2.6 shows the influence of incorrect assumption of the lidar ratio exponent k from Eq. 2.14. While the inversion is certainly sensitive to the lidar ratio exponent, it still strays in this case by less than 40 % at most from the correct value, which is not much considering that values of the extinction coefficient range over several orders of magnitude. Finally,

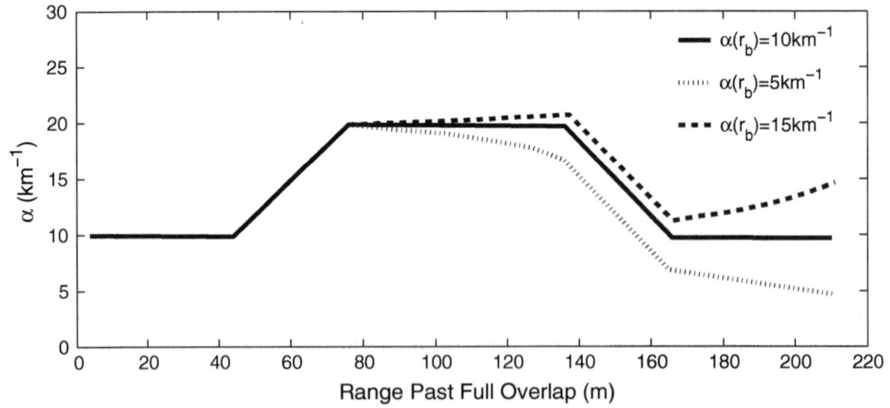

Fig. 2.5 Influence of boundary extinction value $\alpha(r_b)$ on backward inversion as originally calculated by Klett [18]

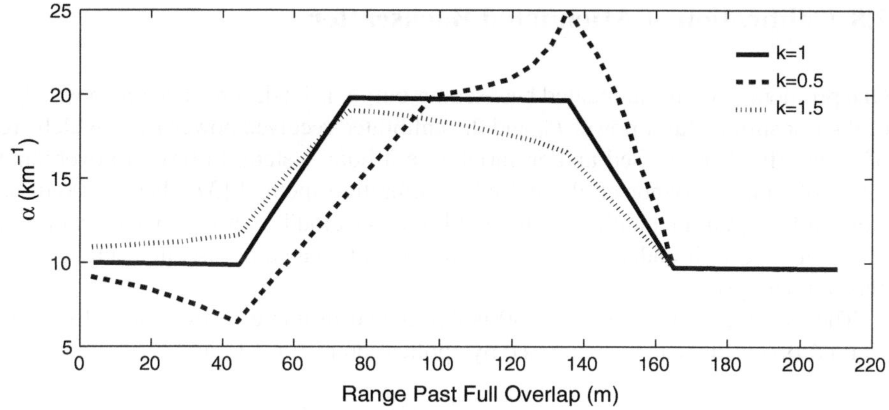

Fig. 2.6 Influence of lidar ratio exponent k on backward inversion as originally calculated by Klett [18]

since range-corrected signals are typically noisy at far ranges, the influence of this noise on inversion also needs to be considered. A thorough examination of error sensitivity of the backward inversion technique has been given by Rocadenbosch and Comerón [22].

2.7 Attenuated Backscatter

Since inverted data depends greatly on the inversion technique, the inverted profile is often not the best profile for inter-comparison of different instruments. The profile used most widely for instrument comparison is the attenuated backscatter, that is, the range-corrected, overlap-corrected, calibrated signal. It is a standard output of satellite lidar data [35] and is also a typical ceilometer output.

Considering discussion in [35, 36], attenuated backscatter $\beta'(r)$ can be expressed, by solving the lidar equation for the product of backscatter and transmission, as

$$\beta'(r) = \frac{P(r)r^2}{K_s O(r)} = \beta(r) T_{\text{tot}}^2(r), \tag{2.30}$$

where T_{tot} is the total transmission through all scatterers. The range-corrected power $P(r)r^2$ divided by the product of the overlap function $O(r)$ and the system constant K_s therefore reflects the combined contributions of backscatter and attenuation. By fully expressing the system constant K_s using Eq. 2.15, the attenuated backscatter becomes

$$\beta'(r) = \frac{P(r)r^2}{P_0 \frac{\tau c}{2} O(r) A_0}, \tag{2.31}$$

given throughout this thesis in units of $\text{m}^{-1}\text{sr}^{-1}$.

2.8 Calibration of Attenuated Backscatter

One problem with the attenuated backscatter output is its dependence on knowledge of the transmitted laser power P_0 and the calibrated received power $P(r)$, which are often poorly characterised and/or variable in a lidar system. In order to overcome this problem, O'Connor et al. devised a calibration method [37] that uses returns from fully signal-attenuating stratocumulus water clouds, whose optical scattering properties are well understood, to calibrate the attenuated backscatter output of an elastic lidar system.

The transmission T_{tot} in Eq. 2.30 is dominated in this case by returns from the cloud droplets and can be expressed by application of Eq. 2.11 as

$$T_{tot}(r) = e^{-\int_0^r \alpha(r')dr'} = e^{-\tau_\alpha}, \tag{2.32}$$

where, as previously noted, the integrated extinction $\int_0^r \alpha(r)dr$ is equivalent to the optical depth, τ_α. Equation 2.30 can therefore be rewritten

$$\beta'(r) = \beta(r)e^{-2\tau_\alpha}. \tag{2.33}$$

Instead of defining the lidar ratio as the backscatter to extinction ratio $\Pi_p = \frac{\beta}{\alpha}$, O'Connor et al. used the extinction to backscatter ratio and defined it as $S = \frac{\alpha}{\beta}$. Assuming an infinitesimal change dr in range, for the single scattering case the corresponding change $d\tau_\alpha$ in optical depth can be expressed as

$$d\tau_\alpha = S(r)\beta(r)dr. \tag{2.34}$$

A factor, $\eta_m(r)$, that corrects for multiple scattering effects and ranges from 0.5 to 1 can then be introduced such that

$$d\tau_\alpha = \eta_m(r)S(r)\beta(r)dr. \tag{2.35}$$

If the attenuated backscatter as stated in Eq. 2.33 is integrated over the entire range of the instrument, the resulting value B is expressed

$$B = \int_0^\infty \beta'(r) = \int_0^\infty \beta(r)e^{-2\tau_\alpha}dr. \tag{2.36}$$

If Eq. 2.35 is substituted into this expression and η_m and S are assumed constant, it becomes

$$B = \int_0^\infty \beta'(r) = \frac{1}{\eta_m S}\int_0^\infty e^{-2\tau_\alpha}d\tau_\alpha. \tag{2.37}$$

Since the integral $\int_0^\infty e^{-2\tau_\alpha} d\tau_\alpha$ can be evaluated such that

$$\int_0^\infty e^{-2\tau_\alpha} d\tau_\alpha = \frac{1}{2}, \tag{2.38}$$

the integrated attenuated backscatter becomes

$$B = \int_0^\infty \beta'(r) = \frac{1}{2\eta_m S}, \tag{2.39}$$

and the attenuated backscatter calibration should be adjusted until this is true.

O'Connor et al. specified that the stratocumulus cloud used for calibration must have a peak backscatter coefficient of greater than 1×10^{-4} sr^{-1} and the signal level at this height must be at least 20 times greater than that 300 m above. In addition, no precipitation or strong aerosol events should be present during calibration. Using a droplet size spectrum width parameter (μ) ranging from 2 to 10, and median droplet diameters ($2 \times a_0$) between 4 and 10 μm to define the droplet size distribution in a thick stratocumulus cloud, they derived an effective lidar ratio of $S = 18.8 \pm 0.8$ sr at 905 nm as the appropriate lidar ratio for this technique. The multiple scattering correction factor η_m is discussed further in Chap. 5 where this method is applied.

This calibration technique provides a useful means of calibrating the signal output from an instrument whose system parameters may be poorly characterised or subject to drift.

2.9 Conclusion

This chapter has discussed fundamental theoretical tools used to describe the nature of scattering particles in clouds and the mechanics of elastic scattering in the atmosphere. It has introduced the single-scattering elastic lidar equation and the parameters it contains, including the geometry involved. It has presented a derivation of the classical backward inversion technique used in elastic lidar and shown its robustness. Finally, it has explained the attenuated backscatter function used for inter-comparison of lidar returns along with a method by which it can be calibrated. The geometry of lidar measurement has largely guided the design of the prototype described in Chap. 3. The lidar equation is applied in a variety of ways in Chaps. 3, 4 and 5, and inversion and calibration both become important in Chap. 5.

References

1. H. Pruppacher, J. Klett, *Microphysics of Clouds and Precipitation* (Kluwer Academic Publishers, Dordrecht, 1997)
2. D. Deirmendjian, *Electromagnetic Scattering on Spherical Polydispersions* (American Elsevier Pub. Co., RAND Corporation, New York, 1969)
3. M. Abramowitz, I. Stegun, *Handbook of Mathematical Functions: with Formulas, Graphs, and Mathematical Tables*, Applied mathematics series (Dover Publications, New York, 1964)
4. A. Kokhanovsky, Optical properties of terrestrial clouds. Earth Sci. Rev. **64**(34), 189–241 (2004)
5. B. Mason, *Clouds, Rain and Rainmaking* (Cambridge University Press, Cambridge, 1975)
6. Q. Han, W.B. Rossow, A.A. Lacis, Near-global survey of effective droplet radii in liquid water clouds using ISCCP data. J. Clim. **7**(4), 465–497 (1994)
7. B. Fomin, I. Mazin, Model for an investigation of radiative transfer in cloudy atmosphere. Atmos. Res. **4748**(0), 127–153 (1998)
8. R. Knollenberg, Clouds, their formation, optical properties, and effects, ser. Techniques for Probing Cloud Microstructure, *Academic Press Rapid Manuscript Reproduction* (Academic Press, San Diego, 1981), pp. 15–92
9. P. Hobbs, A. Deepak, Clouds, their formation, optical properties, and effects, ser. *Academic Press Rapid Manuscript Reproduction*, ed. by P. Hobbs, A. Deepak (Academic Press, San Diego, 1981).
10. D.K. Lynch, K. Sassen, D.O. Starr, G. Stephens (eds.), *Cirrus* (Oxford University Press, New York, 2002)
11. V. Kovalev, W. Eichinger, *Elastic Lidar: Theory, Practice, and Analysis Methods* (John Wiley and Sons Inc, Hoboken, 2004)
12. U. Wandinger, *"Introduction to lidar", in LIDAR: range-resolved optical remote sensing of the atmosphere*, ed. by C. Weitkamp (Springer, New York, 2005), pp. 1–18
13. G. Mie, Beiträge zur optik trüber medien, speziell kolloidaler metallösungen. Ann. Phys. **330**(3), 377–445 (1908)
14. S. Prahl, Mie scattering calculation. (Oregon Medical Laser Center), http://omlc.ogi.edu/calc/mie_calc.html. Accessed July 2012
15. W.J. Smith, *Modern Optical Engineering* (SPIE, New York, 2008)
16. E.W. Eloranta, Practical model for the calculation of multiply scattered lidar returns. Appl. Opt. **37**(12), 2464–2472 (1998)
17. G.M. Hale, M.R. Querry, Optical constants of water in the 200-nm to 200-μm wavelength region. Appl. Opt. **12**(3), 555–563 (1973)
18. J.D. Klett, Stable analytical inversion solution for processing lidar returns. Appl. Opt. **20**(2), 211–220 (1981)
19. F. Rocadenbosch, C. Soriano, A. Comerón, J.-M. Baldasano, Lidar inversion of atmospheric backscatter and extinction-to-backscatter ratios by use of a Kalman filter. Appl. Opt. **38**(15), 3175–3189 (1999)
20. I.S. Stachlewska, C. Ritter, On retrieval of lidar extinction profiles using two-stream and Raman techniques. Atmos. Chem. Phys. **10**(6), 2813–2824 (2010)
21. F. Rocadenbosch, A. Comerón, D. Pineda, Assessment of lidar inversion errors for homogeneous atmospheres. Appl. Opt. **37**(12), 2199–2206 (1998)
22. F. Rocadenbosch, A. Comerón, Error analysis for the lidar backward inversion algorithm. Appl. Opt. **38**(21), 4461–4474 (1999)
23. R. Collis, Lidar: a new atmospheric probe. Q.J.R. Meteorol. Soc. **92**, 220–230 (1966)
24. E.W. Barrett, O. Ben-Dov, Application of the lidar to air pollution measurements. J. Appl. Meteorol. **6**, 500–515 (1967)
25. V.A. Kovalev, H. Moosmüller, Distortion of particulate extinction profiles measured with lidar in a two-component atmosphere. Appl. Opt. **33**(27), 6499–6507 (1994)
26. J.A. Weinman, Derivation of atmospheric extinction profiles and wind speed over the ocean from a satellite-borne lidar. Appl. Opt. **27**(19), 3994–4001 (1988)

27. V.A. Kovalev, Stable near-end solution of the lidar equation for clear atmospheres. Appl. Opt. **42**(3), 585–591 (2003)
28. B. Kaul, Laser sensing the aerosol pollution in the atmosphere, Ph.D. dissertation, Institute of Atmospheric Optics, Tomsk, 1977
29. V. Zuev, G. Zadde, S. Kavkjanov, B. Kaul, Interpretation of Lidar Signals from the Regions of Large Optical Depths, *Remote Sensing of the Atmosphere* (Nauka, Novosibirsk, 1978), pp. 60–68
30. F.G. Fernald, in *proceedings, in Eleventh International Laser Radar Conference* (Madison, Wisc, 1982), p. 213
31. Y. Sasano, H. Nakane, Significance of the extinction/backscatter ratio and the boundary value term in the solution for the two-component lidar equation. Appl. Opt. **23**(1), 11–13 (1984)
32. W. Carnuth, R. Reiter, Cloud extinction profile measurements by lidar using klett's inversion method. Appl. Opt. **25**(17), 2899–2907 (1986)
33. J. Dias, J. Leitao, E. Fonseca, Reconstruction of backscatter and extinction coefficients in lidar: a stochastic filtering approach. IEEE Trans. Geosci. Remote Sens. **42**(2), 443–456 (2004)
34. J.L. Gaumet, J.C. Heinrich, M. Cluzeau, P. Pierrard, J. Prieur, Cloud-base height measurements with a single-pulse erbium-glass laser ceilometer. J. Atmos. Oceanic Technol. **15**(1), 37–45 (1998)
35. M. Vaughan, S. Young, D. Winker, K. Powell, A. Omar, Z. Liu, Y. Hu, C. Hostetler, Fully automated analysis of space-based lidar data: an overview of the CALIPSO retrieval algorithms and data products, in *Laser Radar Techniques for Atmospheric Sensing, Proceedings of SPIE*, vol. 5575, 2004
36. L. Mona, A. Amodeo, G. DâĂŹAmico, G. Pappalardo, First comparisons between CNR-IMAA multi-wavelength Raman lidar measurements and CALIPSO measurements. in *Lidar Technologies, Techniques, and Measurements for Atmospheric Remote Sensing III*, eds. by U. Singh, G. Pappalardo. ser. Proceedings of SPIE-the International Society for Optical Engineering, SPIE, 2007 (Florence, Italy, 2007), pp. 675 010-1–675 010-10
37. E. O'Connor, A. Illingworth, R. Hogan, A technique for autocalibration of cloud lidar. J. Atmos. Oceanic Technol. **21**, 777–786 (2004)

Chapter 3
Opto-mechanical Design of a Biaxial Elastic Lidar Prototype

3.1 Introduction

This chapter describes an eye-safe divided lens biaxial elastic lidar ceilometer prototype, as well as the considerations that were involved in its design. After the initial design specification for the instrument is given, the optical design concept is presented and a method for calculating the optical signal to noise ratio of the instrument is described, as are the key variables involved in optimising it. Several optical configurations are compared, both in terms of optical signal to noise ratio and overlap function, to a leading commercially-available instrument in a similar class. Following this discussion, the final optical design of the prototype is discussed in terms of each of its primary components. In addition, significant factors affecting optical and mechanical tolerancing of the instrument are presented, particularly with regard to sensitivity to changes in temperature.

3.2 Design Specification

Table 3.1 shows the preliminary specification for the prototype instrument that is intended to meet the needs of both the aviation and general meteorological sectors. The range of the instrument needed at minimum to comply with the United States Federal Aviation Administration requirements of 100–12,500 ft (30.48–3,180 m) [1]. However, meteorological stations tend to deploy farther reaching ceilometers, such as the Vaisala CL-31, that have maximum ranges of at least 25,000 ft (7,620 m). The current prototype instrument was designed to extend that range to 10 km in order to put it closer to the next class of ceilometers which includes the Vaisala CL-51 (13 km) and the Jenoptik CHM15k (15 km) described in Chap. 1. The other parameters, with the exception of the planetary boundary layer height report, are fairly standard among commercially available instruments. Note that the field-replaceable laser and detector module requirement places significant constraints on both the

J.D. Vande Hey, *A Novel Lidar Ceilometer*, Springer Theses,
DOI 10.1007/978-3-319-12613-5_3

Table 3.1 Ceilometer design specification

Performance spec.	
Cloud base detection range	30 m–10 km
Sampling range resolution	5 m
Optical range resolution	15 m (for 100 ns laser pulse duration)
Reporting interval	15 s
Simultaneous layer reports	Up to 4 cloud layers
Vertical visibility	Reported if no cloud base reported
Laser safety	Class 1M
Environmental spec.	
Operating temperature	−40 to +60 °C
Operating relative humidity	0–100 %
Entrance window	Self-detect and clear contamination
Solar radiation shield	Protect laser/detector from solar exposure
Additional spec.	
Installation	Size and weight suitable for one person
Replaceable components	Laser/detector modules field-replaceable
Sky condition algorithm	Automatically estimate total cloud cover
Planetary boundary layer	Report PBL height when possible

design and the alignment process. The major item of influence not appearing in the design specification is the cost of manufacture of the unit, which demanded careful consideration but is highly variable due to shifting markets and is omitted due to its commercially-sensitive nature.

3.3 Optical Design Concept

The main performance goal for the instrument was to maximise the far-range optical signal to noise ratio of the instrument while maintaining adequate close-range detection capability. The key manufacturing considerations were to achieve a compact and lightweight design, to minimise the number and cost of optical components, and to limit the complexity and therefore the cost of optical assembly and alignment. Perhaps the most significant constraint on the optical design was compliance with the Class 1M laser safety rating which limits the allowed power density output for a collimated or divergent beam [2]. As is often the case with design, the optimisation process for this prototype involved balancing competing demands of different aspects of the instrument.

Existing lidar ceilometer instruments for general meteorology and aviation applications use biaxial optics (Belfort, Eliasson, AWI, Jenoptik) or common, coaxial optics (Vaisala). The biaxial instruments have the advantage of good optical isolation, while the common optics instrument has the advantage of increased transmitter-receiver overlap at low altitudes. Due to the complexity involved in avoiding optical cross-talk as well as possible commercial issues arising from a Vaisala patent on common optics [3], this approach was not considered. Instead, a novel biaxial system involving a divided aspheric lens was employed in order to combine the advantages of the two standard approaches. Since the transmitter and receiver each has its own optical channel, laser light leakage is prevented and good optical isolation is achieved. Due to close proximity of the optical axes, overlap at low altitudes is greater than that of a full-lens biaxial system with the same laser divergence and receiver field of view. It is noted that early in the design process a similar arrangement with mirrors was considered, but this approach was rejected due to the complexity required in order to minimise partial blocking of the transmitter and receiver apertures by the laser and detector modules.

The prototype lidar ceilometer design that has been developed is a biaxial system utilizing lenses of focal length $f = 335\,\text{mm}$ (at 905 nm), a pulsed diode laser operating at 905 nm, and an avalanche photodiode (APD) detector as shown in Fig. 3.1. The laser beam is isolated from the detector by a dividing wall. This design is innovative in that a single aspheric objective (with a clear aperture diameter of 150 mm) is divided into two elements of semicircular aperture, with optical axes separated by 21 mm. One of the halves is used exclusively in the receiver and the other is used exclusively in the transmitter of this biaxial lidar. The laser is tilted slightly in order to fill the transmitter half lens, but since the laser is placed at the focal point of the lens, the collimated beam leaving the instrument is vertical. The laser and detector are mounted in removable modules on a base plate not shown in this diagram. In addition, each channel includes an optical filter that transmits the laser light but rejects other wavelengths to limit optical noise due to background light and protect the laser and detector components from incidental focused solar radiation.

Besides the relatively large area of the lens apertures, which improves the optical signal to noise ratio, and the close proximity of optical axes, which enhances over-

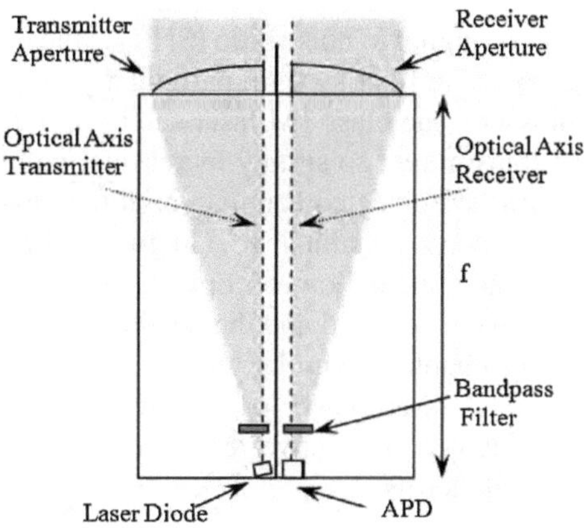

Fig. 3.1 Prototype optical configuration

lap at close ranges, there are a number of reasons for selecting this design. First, the most cost-effective high power pulsed laser sources are laser diodes for which the divergence angle in one axis is typically about twice that of the other due to diffraction at a slit-shaped exit aperture. Because of this, the elliptical beam of the laser approximately fills half of a round lens and therefore uses the transmitter optic efficiently without the need for beam shaping optics if the numerical aperture of the lens is matched to the wider divergence angle of the laser. Secondly, if an appropriate detector area is selected for a given lens focal length, the detector itself acts as the field-stop; this eliminates the need for an additional aperture in the optical system. Another major advantage of this design is the fact that cutting two lenses from one element gives much closer focal length agreement between the two halves than would be expected, due to manufacturing tolerances, between two elements polished individually. Provided the rotational tolerances of the lens specification are acceptable, each half of the divided lens produces a focal point at the same effective distance, making for, effectively, a shared focal plane. This allows the laser and detector to be focused in parallel by translation of the plate on which they are both mounted. Since the modules need to be field-replaceable, they must be pre-focused before being installed in the optical assembly, and therefore this parallel focusing technique can be exploited for ease of alignment. Finally, the divided-lens configuration allows the optical assembly to be compact and the aspheric component itself also offers potential cost savings over typical two-lens optics.

Through efficient use of a small number of optical components, the cost and complexity of the design are minimised without compromising performance. In the following sections, signal to noise ratio, overlap, and the primary optical components and their arrangement are discussed in detail.

3.4 Design Approach for Optimisation of Optical Signal to Noise Ratio

A good signal to noise ratio (SNR) is the key to making sensitive measurements with an eye-safe lidar system, particularly during the daytime when background light is an issue. The class 1M laser safety rating restricts the laser power density that can be transmitted, so simply increasing the laser output power until the desired signal returns are detected is not an option unless the transmitter area is increased and the power density maintained. Gregorio [4] presented a method, recently expanded [5], for evaluating ceilometer optical designs whereby the receiver lens area and field of view were adjusted and the resulting SNR and overlap calculated. The method used in this chapter is similar, though the transmitter aperture area, and correspondingly the total laser power output, were also varied for SNR analysis, and the SNR is evaluated in relative terms compared to a reference instrument rather than in absolute terms.

Assuming that a standard silicon APD detector is used, that a variety of laser powers are available, and that laser divergences can be matched with the numerical apertures of the lenses, the key variables for the optical design are the area of the laser transmitter optic, A_T, the area of the receiver optic, A_0, the bandwidth of the laser, $\Delta_{\lambda T}$, the bandwidth of the optical bandpass filter, $\Delta_{\lambda R}$, the maximum half angle divergence of the laser transmitter, θ_T, and the half angle field of view of the receiver, θ_H. Note that the laser transmitter divergence θ_T was not allowed to exceed the field of view of the receiver θ_H.

For detection of a discrete random process, the signal to noise ratio, SNR, can be expressed as

$$\mathrm{SNR} = \frac{P}{\sqrt{N_B + P}}, \tag{3.1}$$

where P denotes signal counts and N_B denotes noise counts. In most daytime lidar applications, a far-range signal from a target at distance r is dwarfed by background light so that $N_B \gg P(r)$, and the approximation

$$\mathrm{SNR}(r) \approx \frac{P(r)}{\sqrt{N_B}} \tag{3.2}$$

can be applied.

The range-resolved measured signal power, $P(r)$, at a given laser wavelength can be determined using the standard single-scattering elastic lidar equation given in Eq. 2.16. If the system constant from Eq. 2.15 is expressed fully, the lidar equation can be written

$$P(r) = P_0 \frac{c\tau}{2} \eta A_0 \beta(r) \frac{O(r)}{r^2} \left(e^{-2\int_0^r \alpha(r')dr'} \right). \tag{3.3}$$

By the class 1M laser safety standard [2], the cross-sectional laser beam power density L_D in the plane of the transmitter lens for a given pulse length and repetition frequency is restricted, but if the laser beam is collimated, the total laser power is not. Therefore L_D can be fixed and the area, A_T, of the transmitter lens varied accordingly for comparison of optical design SNR. For the purposes of this discussion, the substitution $P_0 = L_D A_T$ is made such that

$$P(r) = L_D A_T \frac{c\tau}{2} \eta A_0 \frac{O(r)}{r^2} \beta(r) \left(e^{-2\int_0^r \alpha(r)dr} \right). \tag{3.4}$$

Here L_D is assumed to be uniform in the plane of the transmitter lens to facilitate simplified comparison of possible optical designs, though in reality the beam is likely to have a Gaussian or approximately Gaussian profile.

The noise due to background light can be found using the sky background radiance equation [6],

$$N_B = f(\lambda)\Omega\Delta_{\lambda R} A_0, \tag{3.5}$$

where $f(\lambda)$ is the radiance in $W/(sr\ nm\ m^2)$ of background light at the centre wavelength of the filter, and Ω is the solid angle describing the receiver field of view.

For a small field of view (FOV) plane angle, the solid angle can be approximated from the full angle receiver FOV, θ_F, by

$$\Omega \approx \frac{\pi {\theta_F}^2}{4}. \tag{3.6}$$

For simplicity, the half-angle receiver FOV is used for all calculations. If θ_H is the half-angle FOV, the solid angle FOV is expressed as

$$\Omega \approx \pi {\theta_H}^2, \tag{3.7}$$

and the sky background radiance can be written in terms of half-angle FOV, θ_H, such that

$$N_B = f(\lambda)\Delta_{\lambda R}\pi {\theta_H}^2 A_0. \tag{3.8}$$

For a uniform laser distribution the daytime optical SNR at range r can be found by substituting Eqs. 3.4 and 3.8 into Eq. 3.2 such that

$$\mathrm{SNR}(r) = \frac{L_D A_T \frac{c\tau}{2} \eta A_0 \frac{O(r)}{r^2} \beta(r) \left(e^{-2\int_0^r \alpha(r)dr}\right)}{\sqrt{f(\lambda)\Delta_{\lambda R}\pi {\theta_H}^2 A_0}}. \tag{3.9}$$

Assuming other parameters such as system efficiency and laser pulse length and repetition rate are equal, that comparison lidar measurements are made at the same time, location, and direction (backscatter and attenuation over the path length are fixed, as is the background light level) and that a comparison range beyond full overlap is used such that $O(r) = 1$, the optical signal to noise ratio can be simplified to

$$\mathrm{SNR}(r) = \frac{A_T\sqrt{A_0}}{\theta_H\sqrt{\Delta_{\lambda R}}} K(r), \tag{3.10}$$

where $K(r)$ is

$$K(r) = \frac{L_D c\tau\eta\beta(r)\left(e^{-2\int_0^r \alpha(r)dr}\right)}{2r^2\sqrt{f(\lambda)}}. \tag{3.11}$$

Equation 3.10 reduces optimisation of signal to noise ratio to four parameters. The SNR for an eye-safe instrument with a fixed laser power output density measuring in fixed atmospheric and background light conditions is directly proportional to the area of the laser transmitter and the square root of the receiver area, and inversely proportional to the field of view angle and the square root of the optical filter bandwidth.

The calculated signal to noise ratios of a number of optical designs are given in Table 3.2. The optical arrangements of these designs fall into four categories.

Table 3.2 Relative SNR of various optical designs

Design	Lens dia. (mm)	Lens configuration	Focal length (mm)	Half-angle laser divergence (mrad)	Half-angle FOV (mrad)	Optical filter FWHM (nm)	SNR
CL31[a]	96	Coaxial 50–50 % by area	300	0.7[b]	0.83	36	1
CLCO[c]	96	Coaxial common optics	300	0.39	0.83	36	0.7
A	96	Biaxial whole lenses	300	0.39	0.83	36	2.8
B	106	Biaxial full lenses	335	0.35	0.75	36	4.3
C	150	Biaxial half lenses	335	0.35	0.75	36	4.3
D	150	Biaxial half lenses	335	0.35	0.75	18	6.0
E	150	Biaxial half lenses	335	0.35	0.45	36	7.1
F	150	Biaxial half lenses	335	0.35	0.45	18	10.0
G[c]	150	Coaxial common optics	335	0.35	0.75	36	8.5
H	150	Biaxial full lenses	335	0.35	0.75	36	12.0

[a] Design information from [7] and [8], lens area sharing assumed to be 50 % per channel, filter bandwidth estimated 36 nm
[b] 0.39 mrad used in overlap calculation as discussed in text
[c] 50 % beamsplitter included (laser power doubled to compensate, receiver lens area multiplied by $\sqrt{2}$ to account for losses)

The first of these is the configuration of the Vaisala CL31 shown in Fig. 3.2. In this configuration a common lens is used for both transmitting and receiving, but a different region of the lens is used for each of these. A ring mirror is used to reflect the outer part of the lens into the detector while the laser beam is transmitted through the hole in the middle. The second configuration is a fully shared common optics system of the type that was used in an older Vaisala model, the CT25k. This configuration, used in designs CLCO and G in Table 3.2, is shown in Fig. 3.3. Figure 3.4 shows the typical dual-lens configuration used in designs A, B, and H, as used by AWI and Eliasson. The fourth design type, shown in Fig. 3.5 is the divided-lens biaxial design that was selected for the prototype design C, and was also used in designs D, E, and F.

A fifth arrangement, based on design C, is called Modified C. In this design the lens is divided in half for full optical isolation, but the gap between the channels is minimised such that it is just big enough for a thin optical barrier. In order to get the two channels as close together as possible, a fibre-coupled laser and detector are used, with a slit aperture on the laser output fibre to match the elliptical diode laser

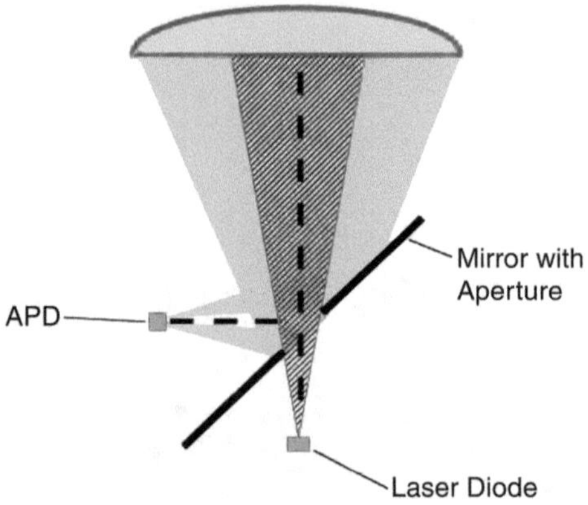

Fig. 3.2 Optical configuration of the Vaisala CL31

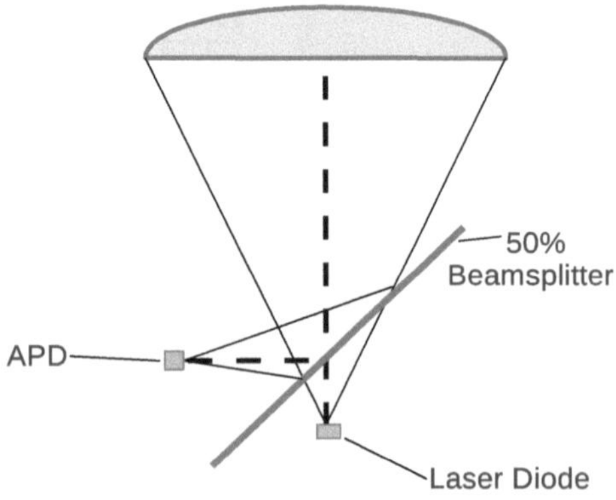

Fig. 3.3 Optical conguration of a fully shared common optics system as used in design CLCO and design G

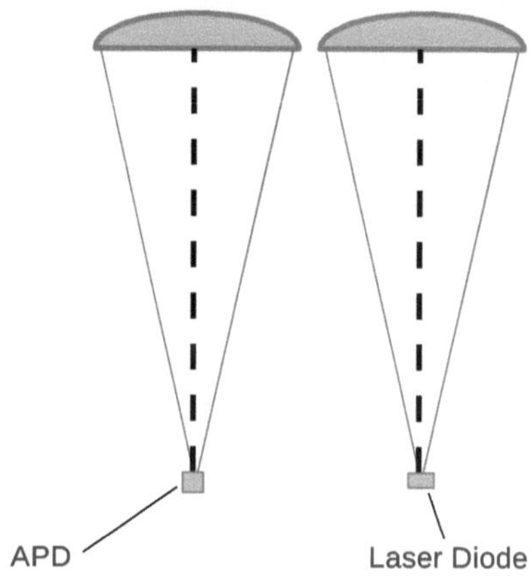

Fig. 3.4 Standard dual-lens biaxial configuration used in designs A, B, and H

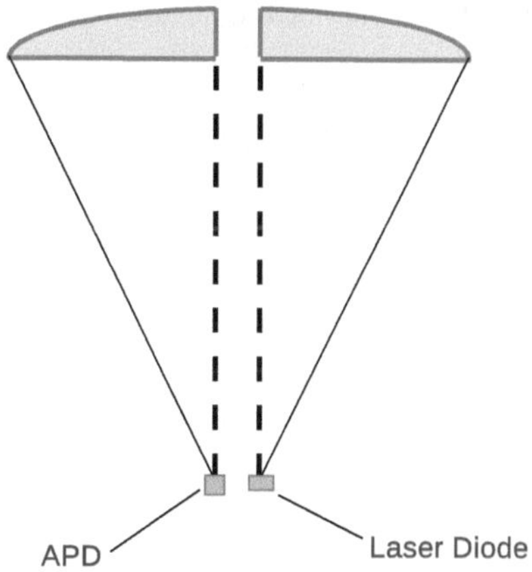

Fig. 3.5 Divided lens configuration with optical axes offset as employed in prototype design C as well as in designs D, E, and F

beam divergence to fill the half lens. This design, shown in Fig. 3.6, has the same SNR as the prototype design C.

As the lidar prototype considered in this work is intended for commercial use, the optical design of a successful commercial instrument, the Vaisala CL31, was used as a reference for optical SNR. The four primary SNR parameters from Eq. 3.10 were used to compare the signal to noise ratios of a number of possible designs to this reference. Table 3.2 shows the relative signal to noise ratios calculated using Eq. 3.10, selecting a $K(r)$ value such that SNR = 1 for the Vaisala CL31. After considering size and cost of lenses, optical, mechanical and manufacturing tolerances, and overlap profile, design C with an SNR factor of 4.3 was selected as the optimal design for this instrument.

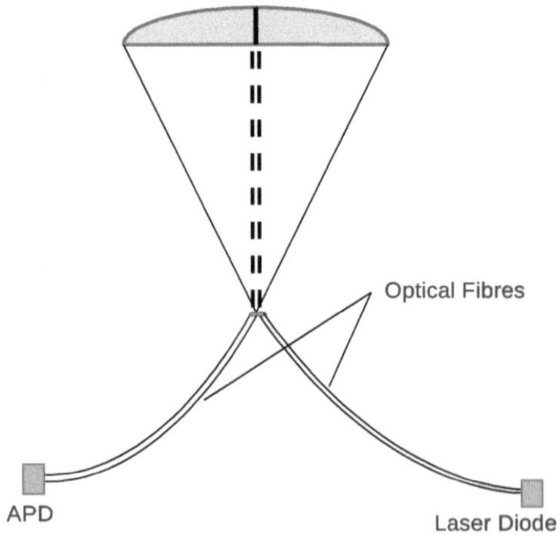

Fig. 3.6 Divided lens configuration of design modified C with minimal offset of optical axes

3.5 Field of View and Overlap Function

By a geometrical process that is described in Chap. 4, the overlap function that determines effective receiver response with range was calculated for the designs from Table 3.2 as shown in Fig. 3.7. For all designs considered, a laser diode laser with an emitter stripe of 0.235 mm was used; the light emitted from this stripe was assumed to be highly multi-mode and therefore incoherent. Also used in all designs was a circular detector of 0.5 mm diameter. The receiver FOV and transmitter divergence were varied only by adjusting focal length. The primary variables considered were

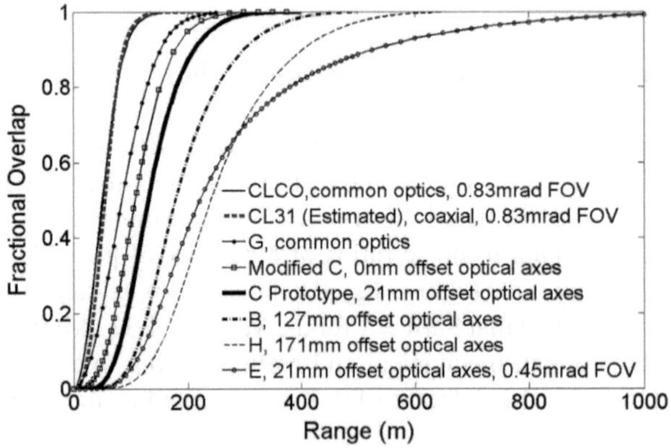

Fig. 3.7 Calculated overlap function for possible lidar prototype optical designs. Half-angle FOV = 0.75 mrad unless otherwise noted

the size, shape, and position of the lenses. For biaxial arrangements besides Modified C, the spacing between the closest edges of the lenses was 21 mm.

Note that since the parameters of the laser and possible beam shaping optics have not been published for the Vaisala CL31, these had to be estimated in its overlap calculation. In an effort to make a direct comparison between optical configurations, the same 0.235 mm laser stripe with no beam shaping optics was assumed for all designs, giving the CL31 a fast-axis laser divergence of 0.39 mrad rather than the specified 0.7 mrad. Since the distribution of the CL31's outgoing laser beam was not known, its overlap was calculated using both uniform and Gaussian distributions, but this was found to have negligible influence on overlap. While a full overlap height of approximately 70 m has been reported for the CL31 in an article by Martucci et al. [9], the estimation here places it closer to 100 m. Should more details of the CL31 design become available, this calculation could be refined accordingly. Note that it is not just the lens arrangement, but also the faster laser divergence and wider field of view that contribute to the fast overlap turn-on of the CL31. Use of a 0.83 mrad half-angle (CL31) rather than a 0.75 mrad (design C) gives its overlap function a greater slope than the other designs as can be seen in Fig. 3.7, but this also reduces the signal to noise ratio by approximately 10 %.

Perhaps the most important conclusion from these calculations is that design B, based on 106 mm diameter circular lenses, each with the same area as half of a 150 mm diameter lens, has the same optical SNR as the prototype design C but takes approximately 150 m farther to reach full overlap. Also of interest are the variations between shared optical axis configurations. First, design CLCO, which is identical to the CL31 except that it has fully shared common optics rather than 50–50 % shared optics, was shown to have an overlap very similar to that of the CL31. Second, it is interesting to note the overlap difference between the 150 mm common optics arrangement G, based on a fully-shared lens, and the divided-lens design, Modified C, with only a thin divider between its two half lenses and therefore essentially no offset between optical axes. Design Modified C takes about 20 m longer to reach full overlap than design G. Though the Modified C design, in turn, reaches full overlap about 20 m sooner than the prototype design C, this difference was not considered large enough to justify the cost and complexity of implementing it.

It was shown in Eq. 3.10 that the optical SNR of the instrument is inversely proportional to the field of view (FOV) angle θ_R. From this perspective, the smallest possible FOV is desirable. Also, a smaller FOV means less multiple scattering contribution to the signal and therefore a more accurate inversion by the single-scattering lidar equation. However, a smaller FOV also means a longer distance to overlap onset and full overlap, which in turn means reduced close-range sensitivity of the instrument. Also, as the detector aperture size (and therefore FOV angle) approaches the laser aperture size (and therefore divergence angle), alignment tolerances become tighter and it becomes increasingly difficult to align the system without wasting signal by failing to focus it onto the detector. In addition, if the detector aperture is larger than the laser aperture, the instrument will be less sensitive to defocus effects for near-range signals. For these reasons the detector aperture was constrained to be no smaller than

0.5 mm diameter, making it approximately 2 times larger than the 0.235 mm laser aperture length and allowing for realistic alignment and focus tolerances.

There is one more consideration that must be made regarding overlap. For good close-range sensitivity, overlap onset is required at the shortest possible distance, but detector saturation due to strong near-range returns should be avoided. To balance these concerns, along with the concerns discussed previously in this section, the overlap profile of the divided-lens design C was deemed to be favourable. Note that the turn-on of its overlap is clearly faster than those of traditional biaxial designs B and H. In order to check the the accuracy of the calculated overlap, the overlap of a prototype instrument was characterised by the measurement method described in Chap. 4. As will be shown, the overlap of the design implemented into an actual instrument agreed throughout most of the profile to within ± 15 m of the calculated overlap and provided a calibration that could allow for reasonable inversion for close range aerosol detection.

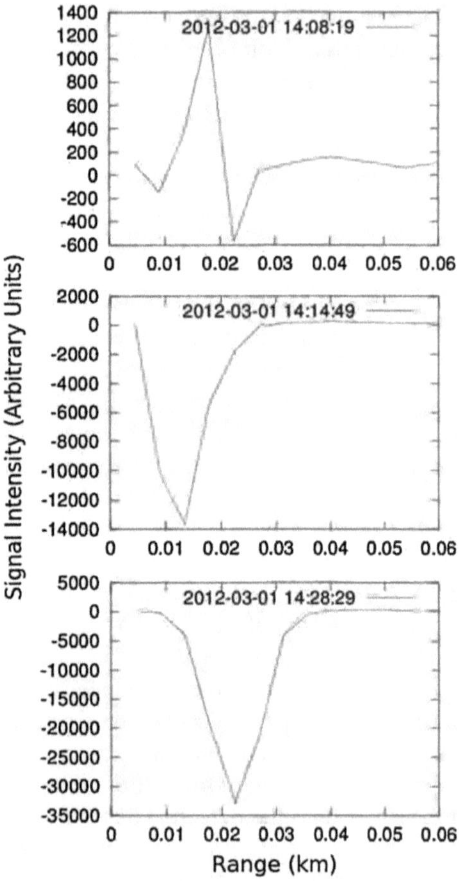

Fig. 3.8 Design C pre-overlap multiple scattering cloud returns (−polarity). *Top* electronic noise signature only. *Centre* cloud at approximately 0.01 km. *Bottom* Cloud at approximately 0.02 km. Courtesy Campbell Scientific, Ltd

It is important to note that overlap onset is not the same as minimum detection range, since multiple scattering of close range aerosols, such as water droplets in fog, makes for significant lidar returns even before overlap onset. The near-range response of the prototype instrument for cloud detection by multiple scattering was tested by aiming the instrument at a cloud created by a steam pressure washer. The results of these measurements are shown in Fig. 3.8 as raw, non-range-corrected signals that illustrate sensitivity to cloud droplets at close range; the top panel shows shows the noise signature without cloud detection, the centre panel shows pre-overlap signal detection from 10 m and the bottom panel shows greater signal detection amplitude from 20 m (0.02 km), where overlap onset increases the detected signal level.

3.6 Laser Considerations

The first consideration for laser selection is the wavelength of the laser. For elastic lidars, the key consideration is atmospheric transmittance of the wavelength. As shown in Fig. 3.9, there are a number of wavelength bands in the spectral region from the ultraviolet through the near-infrared that allow good transmission of light through the atmosphere.

In addition to considering transmittance, the solar spectrum needs to be considered in order to evaluate background light levels at various wavelengths. Figure 3.10 shows the solar spectrum both outside the atmosphere and at sea level, and also indicates the primary absorbing species for each major absorption band. While of course lower transmittance at a given wavelength means less solar radiation at that wavelength, taking the black-body spectrum of solar light into consideration reveals, for example,

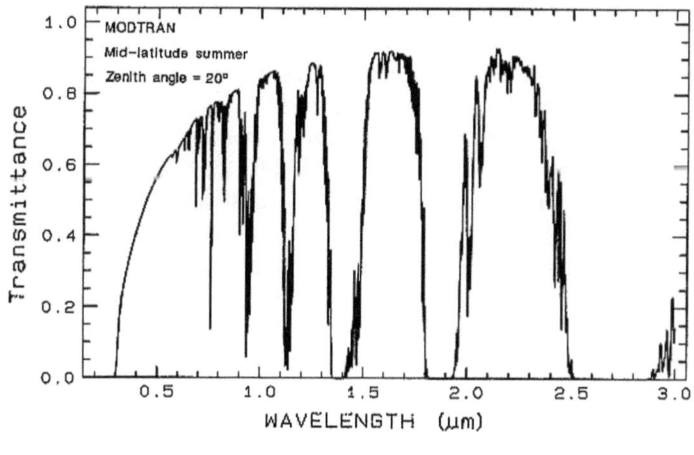

Fig. 3.9 Modelled atmospheric transmittance as a function of wavelength, reproduced from [10] (public domain)

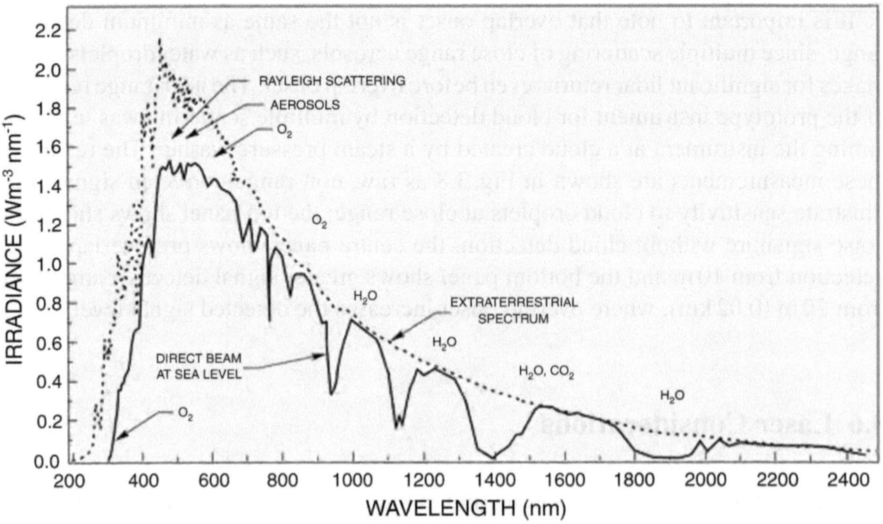

Fig. 3.10 Solar spectrum above the atmosphere and at the Earth's surface, reproduced from [11] (used with permission from Oriel instruments, a Newport Corporation Brand)

that while transmittance at 1,064 and 1,550 nm are similar, both being greater than 0.85, the solar radiation level at 1,550 nm is approximately half of what it is at 1,064 nm.

Since visible laser light can distract pilots or attract unwanted attention to the presence of an expensive meteorological station, lidar ceilometers are typically designed to be invisible to humans. The 1,064 nm NdYAG wavelength, widely used in research lidars due to a high atmospheric transmittance of around 0.85, is also used in high cost ceilometers such as the Jenoptik CHM 15 k, however, this technology is prohibitively expensive for a low cost ceilometer. Due to reasonable atmospheric transmission at 905 nm of approximately 0.6, as well as the wide availability of inexpensive diode lasers at 905 nm and silicon APD detectors sensitive to 905 nm, this is the typical wavelength selected for lidar ceilometers.

Another wavelength, 1,550 nm, is a good candidate for use in invisible wavelength eye-safe lidar. According to Saito et al. [12], because absorption of 1,550 nm light by liquid water in the the lens and cornea of the human eye limits retinal damage, the maximum permissible exposure to this wavelength allowed by laser safety regulations is, by their calculations, approximately five orders of magnitude greater than that allowed near 1 μm. They also emphasised the wide availability of components at this wavelength due to its use in telecommunications applications. As mentioned previously, an additional advantage of 1,550 nm is the fact that the solar background level is about 50 % lower than it is at 905 nm, and besides this it has an atmospheric transmission of around 0.9, 50 % greater than that at 905 nm. However, the same higher absorption by water of 1,550 nm light that makes it "eye-safe", approximately 175 times greater than that of 900 nm light [13], means that the liquid water droplets being measured in the atmosphere would be expected to attenuate a 1,550 nm beam significantly more than a 905 nm beam and therefore lessen the gains somewhat. But due to the much higher laser safety limit and lower background light level at 1,550 nm, there would still be a clear advantage at this wavelength over 905 nm if all else were equal.

Considering the cost constraints placed on the prototype, 1,550 nm technology was found to be prohibitively expensive and not technically viable. As of 2007 when the wavelength for the current prototype was selected, diode lasers at 1,550 nm were found to be approximately 2.5 times the cost of 905 nm diode lasers, which are typically priced in the vicinity of £100 in quantity, and were rated to supply only 60 % of the power of 905 nm diodes. Indium-gallium-arsenide APD detectors at 1,550 nm were approximately 6 times the cost of, had only 15 % as much gain as, and generated significantly more noise than silicon APDs optimised for 905 nm. In order to see optical SNR advantages, laser technology such as fibre or solid state lasers would be needed, and the least expensive fibre lasers were found at the time to start at around £1,500–2,000. For these reasons, 1,550 nm was not pursued. However, if 1,550 nm laser and detector technology becomes available at significantly lower prices, this could provide a major step forward in terms of signal to noise ratio in the future.

Ultraviolet lasers are another possibility for invisible wavelength lidar sources. Eye-safe lidars have been manufactured in the ultraviolet, for example at 355 nm [14]. Atmospheric transmittance here is approximately 0.3, or half of its value at 905 nm, but solar background irradiance is also halved. Lidars at this wavelength again rely on the same expensive Nd:YAG solid-state laser technology that is used at 1,064 nm, but here the 3rd harmonic wavelength is used. Therefore, like 1,550 nm, UV wavelengths do not currently provide sufficient cost advantages to prompt a shift away from 905 nm for this application. After consideration of these options the 905 nm wavelength was chosen, and after some evaluation of various 905 nm diode lasers, a laser with the specification shown in Table 3.3 was selected for the instrument.

Table 3.3 Laser parameters at 21 °C

Peak power	135 W
Centre wavelength	905 ± 5 nm
Spectrum full width half maximum	8 nm
Maximum pulse duration	150 ns
Maximum duty cycle	0.1 %
Divergence parallel to junction plane	12° at 50 % peak intensity
Divergence perpendicular to junction plane	20° at 50 % peak intensity
Drift of centre wavelength with temperature	0.27 nm per °C
Emitter area	235 μm by 200 μm

Since the optical range resolution of a pulsed lidar is considered to be half the distance travelled by light during the pulse duration as discussed in Chap. 2, a laser pulse length of 100 ns was specified in order to achieve the desired optical range resolution of approximately 15 m. To maximise the light output by the laser, the repetition frequency was set to 10 kHz. For a 100 ns pulse duration, a frequency of 10kHz maximises the duty cycle of the laser at its specified limit of 0.1 %. At this repetition frequency a pulse propagates 32 km before the next pulse is sent, which means, taking into consideration the round trip nature of light in a lidar measurement, that the theoretical range limit of the instrument is 16 km. Applying the laser safety guidelines given in the class 1M standard [2], in any 7 mm diameter aperture (typical human pupil size in darkness) the maximum average instrument power output density at 910 nm with 100 ns pulse length and 10 kHz repetition frequency may not exceed 0.298 mW. This is the fundamental optical constraint on the instrument.

When this safety limit is reached, the total power measured outside the prototype (outside the environmental enclosure window) during each pulse is 30 W, which means a pulse energy of 3 μJ per pulse. Before the laser beam leaves the instrument, however, its intensity is reduced by a number of optical effects that must be taken into consideration in order to achieve the desired output level. The external window, AR-coated on the inside but not on the outside, is expected to attenuate the power by 5 %, and the AR-coated lenses have 1 % reflectivity per side. In addition, the optical filter used to shield the laser from direct solar exposure attenuates the power by up to 20 %. Finally, since the half-lenses are designed with an F/# of 2.233 (numerical aperture of 0.216) the angular acceptance along the full angle is 25° and along the half angle is 12.5°. This means that the laser divergence (at 50 % intensity) of 20° by 12° falls within the acceptance angles of the lenses. However, due to the broad, approximately Gaussian, angular power distribution of the laser, 50 % of the laser power is transmitted, while 50 % falls outside of the lens aperture. The accumulation of these losses means that in order to achieve the desired 30 W output pulse power, the laser must be run at approximately 85 W per pulse.

Since beam-shaping optics and prohibitively expensive large lenses were to be avoided in this design, clipping the beam at the 50 % intensity points of the angular distribution is helpful for achieving greater total power out of a lens of a given size. This is because the safety rules limit the power of the maximum intensity point rather than the total power output. Figure 3.11 shows this effect. The solid line shows the power distribution clipped at 50 % of the peak, and the dashed line shows it clipped at 1 %. Since the peak power is limited to the same value for any laser distribution, calculating the area under the curves shows that for this case, collecting close to 100 % of the laser light would reduce the transmitted laser power by half due to eye-safety rules. Thus excluding half of the laser light creates a more even power distribution and approximately doubles the amount of light that can be transmitted by the instrument.

In order for the laser to maintain 85 W pulses reliably at temperatures of 40–60 °C, a 135 W laser was selected. Since this laser has two stacks of emitters, the estimated lifetime of this laser running significantly below its maximum power output

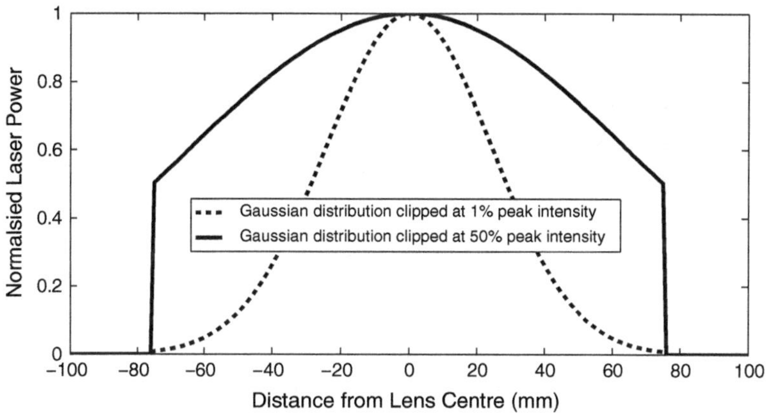

Fig. 3.11 Comparison of power output for different laser distributions though a one-dimensional aperture

is much longer than that of a single emitter stack running at full power. Laser mean time to failure, MTTF, is calculated in years of operation by the manufacturer as follows [15],

$$\text{MTTF} = \frac{1.7 \times 10^{21}}{365 \times 24} \left(\frac{P_0}{L}\right)^{-6} \tau^{-2} F_R{}^{-1} f(T), \tag{3.12}$$

where P_0 is the pulse power in watts, L is the emitter length in μm (235 μm for the lasers considered), τ is the pulse width in ns (100 ns for this system), F_R is the pulse repetition frequency in kHz (10 kHz for this system), and the function $f(T)$ is a multiplier based on laser temperature, shown in Fig. 3.12. From Eq. 3.12 and Fig. 3.12 it is clear that for a given emitter length, pulse length, and pulse repetition frequency, the calculated mean time to failure of the laser reduces significantly with increases in either power or temperature. By application of the formula in Eq. 3.12, a double emitter stack 135 W laser running at 85 W at 50 °C has a calculated mean time to failure of 19 years, whereas for a 70 W single emitter stack laser running at 70 W at 50 °C this is approximately 1 year.

Since the centre wavelength of the laser drifts with temperature, limiting the range of temperatures of the laser reduces the amount of possible drift. Given that the instrument is not hermetically sealed, cooling of the laser was not considered due to concerns over possible condensation. Instead of a cooling or a combined heating/cooling system, a simple resistive heating system was used to keep the laser at 40 °C. Since the hottest temperature expected inside the instrument is 60 °C, the laser temperature is kept within a 20 °C range and the centre wavelength of the laser can drift no more than 5.4 nm. In the middle of this range at 50 °C, the centre wavelength of the laser is 912 nm.

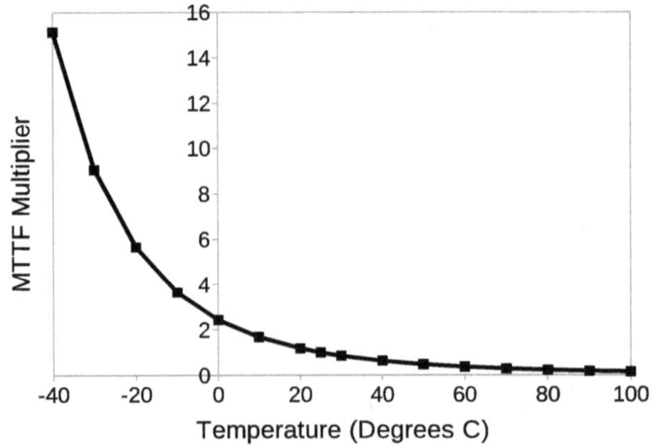

Fig. 3.12 Temperature-dependent multiplier $f(T)$ contribution to MTTF given by [15]

3.7 Detector Considerations

Due to reasons of cost and stability, photomultiplier tubes were not considered for this instrument. Because of its very good performance and stability at a low cost, an APD was the only detector type that was seriously considered. While APDs optimised for photon-counting are available, the count rate of these devices is limited, due to recovery times of around 100 ns, to approximately 10 MHz; analogue-mode APDs, on the other hand, can handle bandwidths in the GHz range [16]. APDs optimised for analogue use are therefore the better choice for ceilometer measurements made under high background light conditions. An analogue-mode silicon APD optimised for high responsivity near 900 nm was selected for the prototype as specified in Table 3.4.

The gain of an APD increases approximately exponentially with increased reverse bias voltage up to its breakdown voltage where it becomes unstable. Since APD gain is quite sensitive to temperature, the performance of the selected APD was characterised at a number of temperatures to simulate possible drift due to changes in the prototype temperature over its required operating range. This was done by supplying the APD with a fixed optical input and varying the reverse bias voltage at

Table 3.4 APD parameters

Detection area geometry	0.5 mm diameter circle
Peak responsivity	860 nm
Spectral response range	440–1,100 nm
Responsivity at gain of 1	0.52 A/W
Quantum efficiency at gain of 1	72 %
Typical gain in instrument	100

each of five temperatures ranging from 30 to 70 °C. Both the gain and the signal to noise ratio were measured at each point. Because gain cannot be measured directly, a gain of 100 was assigned to the APD at a standard operating point of 30 °C and 270 V reverse bias as approximated from the specification. The results of the gain measurement are shown in Fig. 3.13. Note that for 30 and 40 °C the amplifier was saturated beyond the points shown.

These measurements show the expected APD behaviour of exponential increase with reverse bias voltage. They also reveal that beyond 50 °C the maximum gain achievable before breakdown voltage is significantly lower than that reached at lower temperatures. Since the instrument is rarely expected to operate beyond 50 °C, and since the APD is still functional at these greater temperatures, this is not of great concern.

In order to evaluate the noise level of the APD for these data, the signal to noise ratio was determined as shown in Fig. 3.14. Interestingly, the maximum signal to noise ratio of the APD does not occur at the point of maximum gain. The maximum SNR of about 200 turns out to be easily achievable at up to 50 °C, and only decreases by about 25 % at 60 °C compared with the gain which drops off more dramatically.

Since a variable gain amplifier can be used to boost gain without compromising signal to noise ratio, the SNR of the APD is the key parameter for optimisation. From these measurements it can be concluded that keeping the temperature of the APD below 50 °C would be helpful but that the instrument can run at higher temperatures without a prohibitive reduction in performance. Note that in practise a cowling around the internal enclosure provides shade and allows for airflow, thereby helping to reduce the maximum daytime temperature of the enclosed optical system.

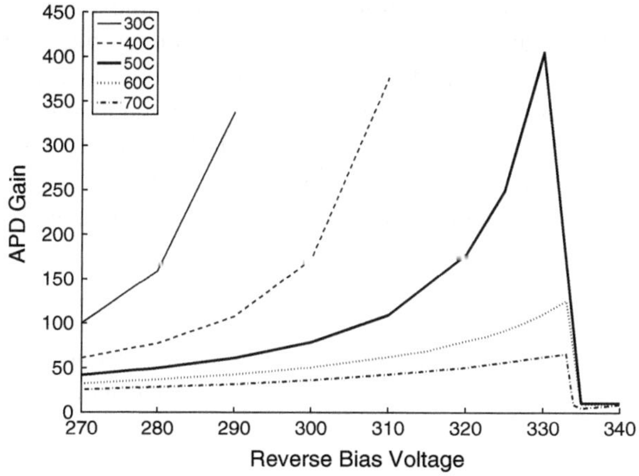

Fig. 3.13 APD gain as a function of reverse bias voltage at five different temperatures

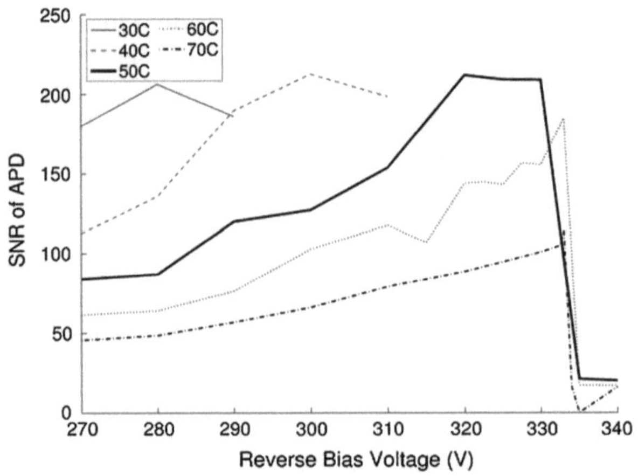

Fig. 3.14 APD signal to noise ratio as a function of reverse bias voltage at five different temperatures

3.8 Optical Bandpass Filter Specification

An optical bandpass filter is used to minimise the amount of background light outside of the laser spectrum that reaches the detector. The filter plays a second role, which is to prevent direct sunlight from damaging the laser or detector should the instrument be pointed incorrectly. For this reason, identical filters are placed in front of the laser and the detector.

The optical filter bandwidth specification depends on the width of the laser spectrum, the laser's centre wavelength tolerance over the whole range of operating temperatures, the range of angles of incidence, filter manufacturing tolerances and the required transmission and attenuation. Based on these parameters, a custom dielectric Fabry-Perot interference filter has been specified as shown in Table 3.5.

Given the laser spectrum of centre wavelength 912 ± 5 nm at 50 °C with full width at half maximum (FWHM) of 8 nm, the filter was designed such that the FWHM of the laser spectrum is contained within the FWHM of the bandpass region. Table 3.6

Table 3.5 Optical bandpass filter manufacturing specification

Parameter	Specification
Filter centre wavelength at normal incidence	915 ± 4 nm
Filter centre wavelength at half of maximum angle of incidence of converging beam	912 ± 4 nm
Filter FWHM	36 ± 4 nm
Transmission in bandpass region	>80 %
Average solar blocking outside bandpass X-ray to 1,600 nm	>OD3

Table 3.6 Optical bandpass filter width contributions

Parameter	Contribution (nm)	λ Min (nm)	λ Max (nm)
Laser centre wavelength at 50 °C	912±5	907	917
Laser centre wavelength shift for ±10 °C	±3	904	920
Laser 8 nm FWHM spectrum bandwidth	−4, +4	900	924
Manufacturer filter centre λ tolerance	±4	896	928
Manufacturer filter FWHM tolerance	−2, +2	894	930
Angular distribution correction	+3	897	933

shows how each of the factors affecting the bandpass filter spectrum is included to produce the specification for the filter. Note that though the default laser temperature in the instrument is 40 °C, it can under extreme conditions reach as high as 60 °C. Because of this the filter is specified for 912 nm, the centre wavelength of the laser at 50 °C. Note also that since wavelength shift of the filter with temperature is an order of magnitude smaller than that of the laser, it is inconsequential over this range of temperatures.

The centre wavelength of an interference filter decreases with increasing angle of incidence. Since angles from 0 to 12.5° are to be accepted by the filter, this shift must be considered. Following a filter bandpass shift calculation given by Andover Corporation [17], an effective centre wavelength shift of approximately −3 nm was determined. Therefore 3 nm was added to the 912 nm laser centre wavelength in order to establish a 915 nm centre wavelength of the filter at normal incidence to be specified to the filter manufacturer.

The additive effects of all of the sources of filter width specification and tolerances are shown in Table 3.6. This table illustrates how each parameter pushes out the maximum bandpass turn-on and turn-off points. Note that a 32 nm bandwidth is required prior to consideration of filter bandwidth manufacturing tolerances of ±4 nm. Since this tolerance means that the width of the filter could be 4 nm narrower than specified, the specified width must therefore be 36 nm to ensure that the FWHM spectrum of the laser always stays within the FWHM bandpass region of the filter.

As shown earlier in this chapter in Eq. 3.10, the signal to noise ratio of the optical design is inversely proportional to the square root of the filter bandwidth. In order to further improve the SNR, filter bandwidth could be decreased by reducing the laser centre wavelength tolerance through temperature tuning or wavelength stabilisation, by selecting a laser with a narrower spectrum or narrowing the laser spectrum by using a narrower filter after the laser, or by collimating the receiver beam and tightening the angle-dependent filter width tolerance. All of these would add cost or complexity, with the possible exception of using a narrower filter in front of the laser. The problem with this approach is that considerably higher laser power would be required and therefore either a more powerful laser would be needed or laser longevity would be reduced.

One essential feature of the filter is its ability to protect the optoelectronic components from direct focused sunlight. After considering various possibilities, including a broadband reflective coating on the lenses, it was determined that the simplest and most cost effective way to protect the laser and detector from direct focused sunlight was to place a bandpass filter with an extended blocking region in front of each component.

BK7, the glass used for the lenses, transmits light from 330 to 2,100 nm. Using numerical integrals of the solar irradiance data provided by the American Society for Testing and Materials [18] over that region, it was determined that 858 W/m^2 could be transmitted by BK7 in direct light. For each of the half lenses the clear aperture area is approximately 0.01 m^2, putting the potential focused power at 8.58 W. If this energy were focused to an area of 1 mm^2, the power density would be 8.58 MW/m^2. The fact that different wavelengths of light focus at different points (for example, 400 nm light focuses at 320 mm, 905 nm focuses at 334 mm, and 1,800 nm focuses at 342 mm) helps to further reduce the power density incident on the detector or laser. To protect the components from potentially damaging radiation, the bandpass filter was designed with an optical density or OD of 3 (10^3 reduction) over a blocking region from UV to 1,600 nm while still maintaining a high transmission of up to 85 % in the bandpass region.

3.9 Lens Considerations

In order to meet the performance requirements of the instrument using only one lens per channel without corrective optics, there are two options, a doublet or an aspheric lens. In order for the lens to accept slightly more than the 20° 50 % intensity angular distribution of the laser, an F/# of approximately 2 is required. As discussed previously, because of safety limitations on the laser power density transmitted by the instrument, a large optical area on the transmitter lens is required in order to obtain a good optical signal to noise ratio. Increasing the area of the receiver lens also improves the SNR. A lens diameter of 160 mm (clear aperture is 150 mm) was chosen as a large yet affordable and manageable size. Since a doublet with $F/\# = 2$ tends to be extremely thick and heavy, as well as expensive, two types of aspheric lenses were tested. First, an inexpensive fire-polished aspheric lens was purchased off the shelf. Second, a custom design for aspheric lenses was generated and optimised using OSLO ray-tracing software [19], and manufactured to specification. This lens was designed to produce a geometric spot radius of 5 μm at focus for a wavelength of 905 nm.

3.9.1 Characterisation of Aspherical Optics

Two aspheric lenses from different suppliers were characterised based on the results of surface form measurements. This section presents the results of this analysis. Note that due to the time-intensive nature of aspheric surface profile measurements, only one lens from each supplier was measured.

3.9.1.1 Characterisation of Fire-Polished Aspheric Lens

The fire-polished condenser lens tested is OEM lens AOI114 from Align Optics Incorporated, based in Florida, USA. This lens was used for the lidar test prototypes, with the lenses cut in half along a diameter. This lens has an effective focal length of 335 mm and a diameter of 160 mm. It has one aspheric side and one spherical side with a large radius of curvature.

A whole, uncoated lens was used for characterization. Using a Taylor Hobson Talysurf PGI 1250 (now called Talysurf PGI Dimension) aspheric form measurement system [20], stylus-based contact surface measurements were made of both sides of the lens. After the best-fitting surface formula for the aspheric side and the best-fitting radius of curvature for the spherical side were determined by the Taylor Hobson analysis software, the form information was entered into OSLO for ray tracing analysis.

With the clear aperture of the modelled lens set at 78 mm and the wavelength set to 905 nm, ray tracing indicated that the spot radius at focus would be 0.2878 mm (about 100 times the diffraction limit) at a focal distance of 330.35 mm from the back surface of the optic. An additional measurement of the aspheric surface was made with the lens rotated 90° from its original position. When analysed by the same method the data indicated a spot radius of 0.2772 mm at a focal distance of 330.56 mm.

In order to study the effect of this minimum spot size limit, the image of the laser stripe was convolved, by a method similar to that detailed in Chap. 4, with the resolvable spot size of the lens to account for the blurring of the laser image by the transmitter lens. This result was then convolved with the same minimum spot size to account for the blurring effects of the receiver lens.

The fractional power of the doubly-convolved laser image collected as a function of detector radius was then calculated as shown in Fig. 3.15. From this it can be seen that only about 50 % of the laser energy that could be detected is collected by a detector with radius 0.25 mm. Because of this, the fire-polished asphere would reduce the potential signal to noise ratio by a factor of at least 2. In addition, the distortion of this lens caused it to produce more than one focal point, making it difficult to judge where the optimum focus location was.

3.9.1.2 Characterisation of Custom Aspheric Lens

A plano-aspheric custom lens design was optimized in OSLO and manufactured to the specification given in Table 3.7. The full details of the lens design are not included because this is commercially-sensitive information. Note that lens manufacturing form tolerances translate to a focal length tolerance of ±2 % or ±6.7 mm which means the mechanical design needs to accommodate this amount of focus variation.

The planar side of the lens was evaluated by placing it onto an optical flat under a sodium lamp. Eight uniform fringes were measured, indicating a slight spherical sag of around 2 μm, which is well within the specified tolerance.

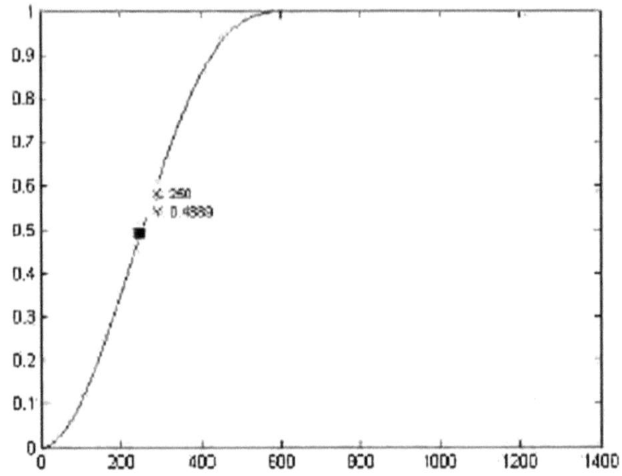

Fig. 3.15 Fractional power collected as a function of detector radius (in μm) using fire-polished aspheric lenses

Table 3.7 Custom asphere specification

Lens diameter	160 mm
Lens clear aperture	150 mm
Lens type	Plano-convex (Aspheric)
Effective focal length	335 mm ± 2 % (6.7 mm)
Resolvable spot radius	5 μm
Maximum surface form deviation	25 μm
Anti-reflective coating	≤1 % reflectivity per surface

The aspheric side of half of this lens was measured on the Taylor Hobson asphere characterisation machine. Six measurements across the lens were made starting from near the centre of the lens at 30° rotational increments, with the lens rotated around the approximate centre of rotation of the surface. Because this lens was already cut in half when it came from the manufacturer and the Taylor Hobson analysis software was not designed to be used on half-lens (it relies on finding the crest of the lens as a reference point), the raw data was entered into MATLAB [21], smoothed, and interpolated. It was then subtracted from surface data generated from the design specification. The position of the measured lens data was adjusted until it had the best fit with the specified data, and the theoretical base radius of curvature of the aspheric surface was also adjusted to find the best fit. The minimized difference is shown in Fig. 3.16.

The maximum surface deviation found in this analysis was 35 μm. The deviation was slowly varying rotationally, which suggests that the MATLAB script did not fully correct for the displacement of the centre of the optical surface of the lens from the centre of rotation of the measurement, or that there was an unknown tilt of

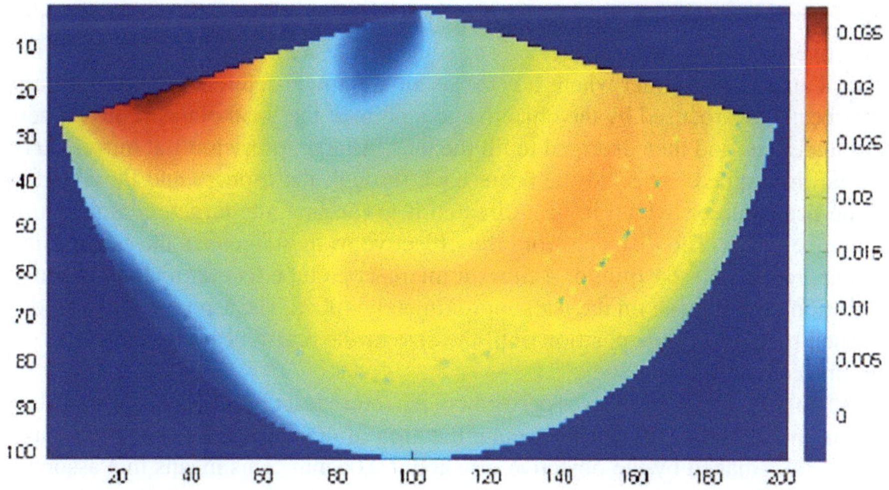

Fig. 3.16 Interpolated minimized surface difference of measured and designed surfaces for custom aspheric half-lens (units in mm)

the lens. Since it was estimated that these possible errors would have increased the amount of apparent surface deviation, the manufactured surface form was deemed to be close enough to the 25 μm specified form deviation tolerance.

In order to double-check the surface quality measurements with the optical performance of the lens, the following verification test was performed. As shown in Fig. 3.17, a 633 nm laser beam was directed through a beamsplitter to a 100×

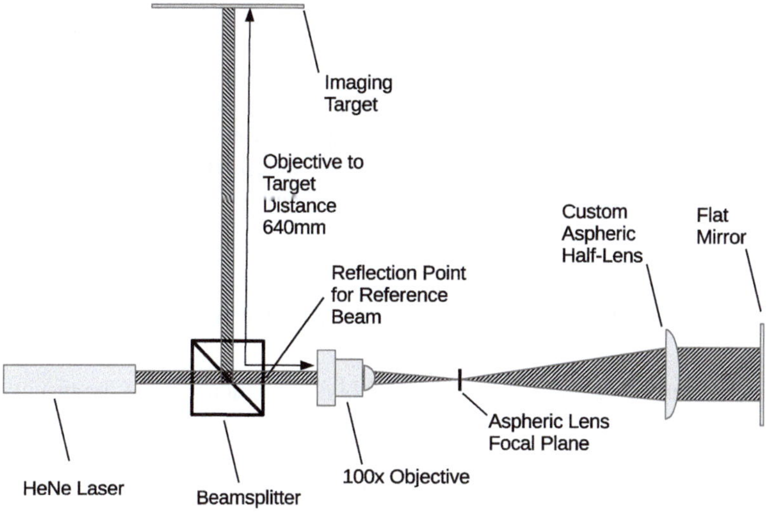

Fig. 3.17 Aspheric quality optical measurement setup

objective. The small reflection of the laser beam off of the glass-air interface on the objective side of the beamsplitter was directed by the semi-reflecting angled surface toward an imaging target where it acted as an alignment reference. The continuing laser beam was focused by the objective to a point at the focal plane of the aspheric lens under test and then diverged to fill the lens. A large, precision-flat mirror ($\lambda/10$) behind the asphere reflected the beam back through the asphere and the objective to the beamsplitter, which in turn directed it to the imaging target, centred on the reference beam. The distance from the objective to the target via the beamsplitter was approximately 640 mm or 4 times a microscope tube length, making the magnification of the image on the wall approximately 400×. The aspheric lens could be focused by adjusting its position until the size of the return beam on the target was minimised.

When this measurement was performed, the return laser light had a radius of about 20 mm in diameter, which with a magnification of 400× suggests that the focused spot radius imaged by the objective was about 0.05 mm. This means that assuming negligible aberration by the components in the optical test system and good optical alignment, a 0.05 mm minimum spot radius is formed by light travelling twice through the aspheric lens, giving the lens the capability of focusing a collimated beam to a spot of 0.025 mm radius. This is roughly 4 times larger than the 0.0062 mm radius calculated from the ideal surface data for 633 nm. Back at the design-optimised wavelength of 905 nm where the ideal minimum spot radius is 0.005 mm, this factor of four times would give an actual minimum spot radius of 0.02 mm. By applying the same convolution technique used for the fire-polished lens (the results of which were shown in Fig. 3.15), even with a 0.03 mm spot minimum radius of the lenses, the slightly blurred image of laser emitter projected by the transmitter lens and reflected back to the receiver from sufficiently large ranges would appear to be about 0.18 mm radius after its trip through the receiver lens and would therefore fit easily onto a 0.25 mm radius detector.

Based on the characterisation measurements presented, the fire-polished lenses were found to be unsatisfactory due to overfilling of the 0.5 mm diameter detector aperture by the blurred image of the laser stripe. The custom-designed lenses, however, were shown to be appropriate for this application, since the slightly blurred image of the backscattered laser stripe can be contained by the 0.5 mm diameter detector aperture used in the instrument.

3.10 Mechanical Design of the Optical Assembly

The mechanical design of the optical assembly is a critical part of the prototype instrument design. It must be adjustable for initial alignment and then fully securable for long term stability. To meet an essential requirement of the design specification it must also accommodate laser and detector module replacement without realignment. After careful consideration of various structural arrangements, a sturdy, easy to assemble design was developed.

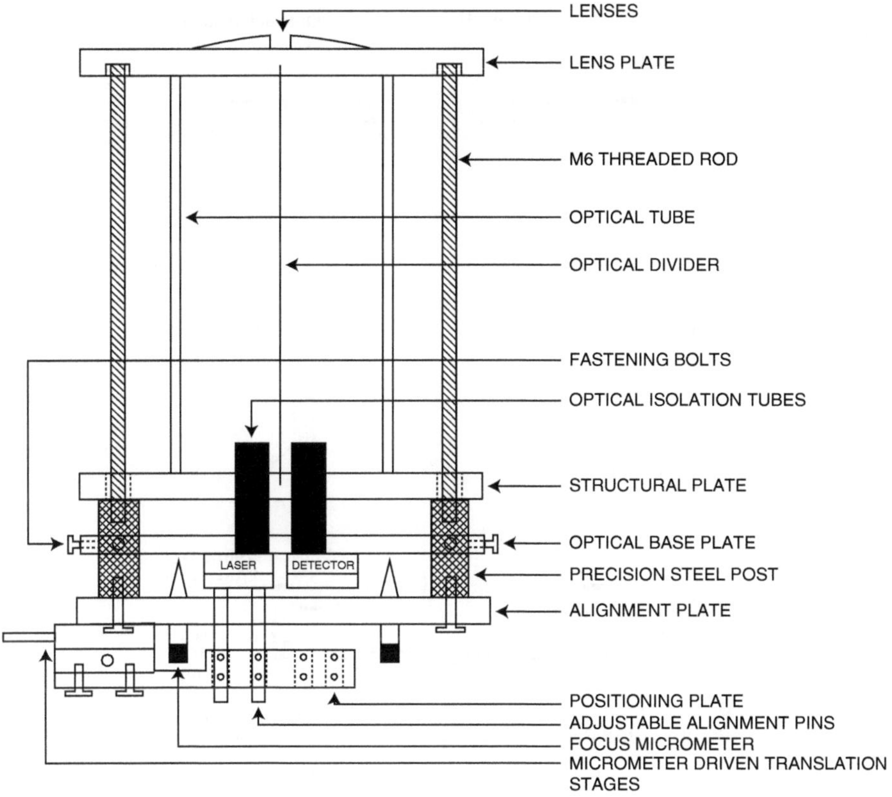

Fig. 3.18 Initial mechanical design concept of optical assembly with alignment tools attached

The initial mechanical design concept for the optical assembly and alignment tooling is shown in Fig. 3.18. An aluminium optical tube is compressed between two aluminium plates by threaded rods that are tightened at the bottom into precision machined steel posts. This provides a stable structure on which to mount the lenses at the top, in recessed pockets, and the laser and detector modules at the bottom, on an adjustable optical base plate that slides along the posts during alignment. An alignment plate with positioning tooling is temporarily bolted onto the bottom of the steel posts during alignment until the modules and optical base plate have been secured. This alignment plate has three focus micrometers positioned in a circle around the plate to raise, lower, and level the optical base plate. When the focus has been set and the optical base plate secured, an X-Y translation stage is used to position the laser and detector modules via height-adjustable alignment pins that accommodate focal length differences from lens to lens. Oversized bolt holes in the module plate allow room for X-Y adjustment.

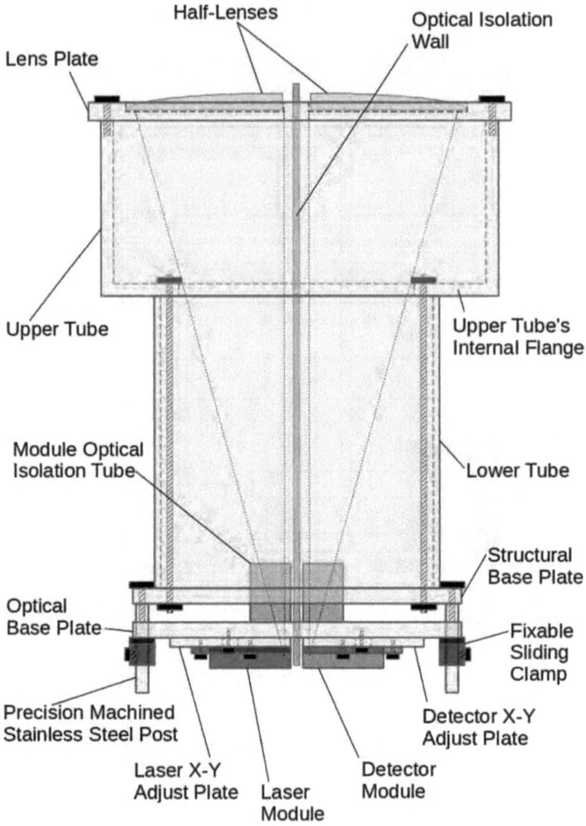

Fig. 3.19 Schematic of refined mechanical design of optical assembly

Several changes to the mechanical design of the optical assembly took place during the process of building pre-production prototypes. With help from the product design team, a lighter, more compact design was created. A schematic representation of this design is shown in Fig. 3.19. The major change was the implementation of a two-part optical tube. The convergence of the beams as they travel down towards focus was exploited to reduce the diameter, and therefore the size and weight of the lower tube. Threaded rods were again used for compression, but here they are kept inside the tube and out of the way. The upper tube is machined with a flange to bolt through at the bottom and has tapped holes in the top to which the lens plate can be directly bolted. The tube sections and the plates were all made from aluminium.

3.11 Optical Alignment

To achieve consistent performance from one instrument to the next, an optical alignment technique capable of focusing the modules to within ±0.1 mm is required in order to keep the instrument within a reasonable range of focus over the range of temperatures it could experience in deployment. And since the laser and detector modules must be replaceable in the field without realignment of the optics, these modules must be manufactured to a high degree of repeatability. These two constraints led to the development of the optical alignment procedure outlined in this section. Note that since the optical design is based on lenses, not mirrors, the alignment technique also required a means of aligning the system at the invisible 905 nm wavelength.

In order to satisfy the field-replaceable module requirement, reference laser and detector modules are needed for use in alignment of each optical assembly. Laser and detector modules are pre-aligned to the reference modules, and all of the optical assemblies are aligned using these same reference modules. In the actual laser and detector modules, the laser emitter and detector aperture are referenced in their respective module plates to the centres of two machined dowel registration holes. These holes mate precisely to two machined alignment dowel pins in the X-Y adjust plate of the optical assembly as shown in Fig. 3.20.

The reference module plates are largely the same as the laser and detector module plates, except that in the reference plates the laser or detector apertures are replaced

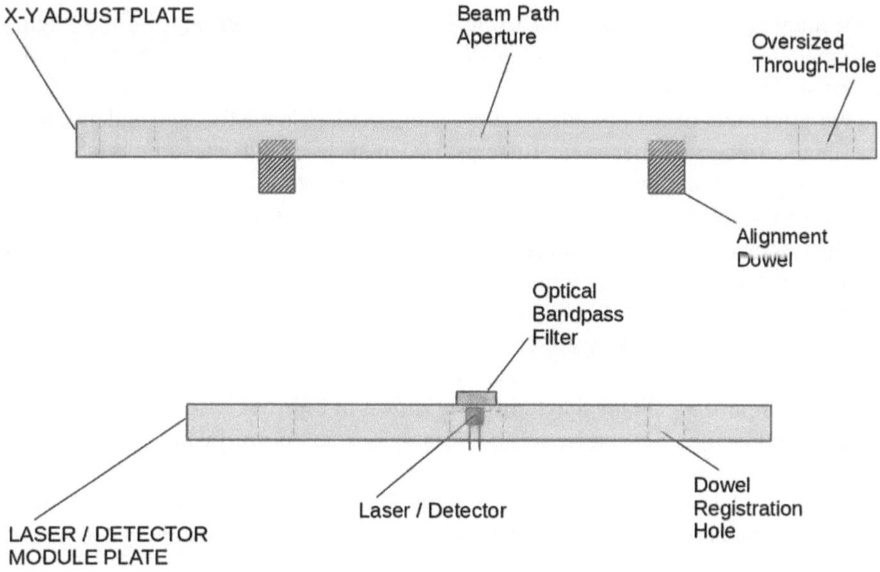

Fig. 3.20 Schematic of laser/detector module plate as it connects to the X-Y adjust plate

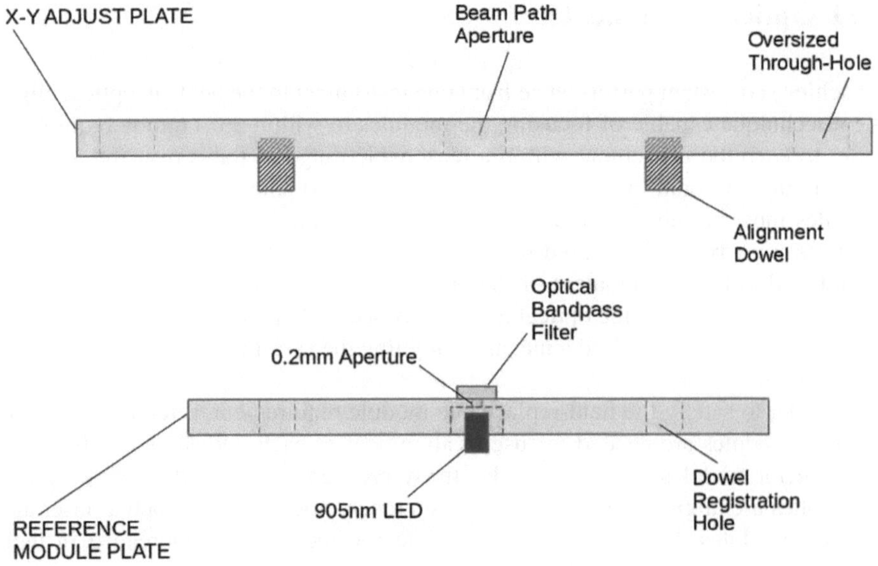

Fig. 3.21 Schematic of reference module plate as it connects to the X-Y adjust plate

by a 0.2 mm diameter aperture which is referenced to the machined dowel registration holes as shown in Fig. 3.21. The optical filters are included on the reference plates in order to ensure that the effective focal distance of the system is the same for both the reference modules and the actual modules. If they were left off the reference plates, a mechanical offset would need to be introduced to compensate for the slightly extended focal distance that occurs due to the high refractive index of the filter material.

Pre-alignment of the modules is done using a microscope focused and aligned to the standard modules, mechanically registered to the laser and detector modules using the same precision dowels arrangement used in the optical assembly. The laser and detector are moved into place with a three-orthogonal-axis plus tilt translation stage, and then they are glued with a proprietary adhesive with high thermal conductivity. Correct laser tilt is checked by imaging the laser beam onto a target before it is glued in place.

The optical assembly itself can be aligned by the procedure summarised in Table 3.8. A collimator was constructed using a 12-inch Newtonian telescope mounted in a custom-built mechanical assembly as shown in Fig. 3.22. A close-up of the optical assembly mounted under the collimator is also shown in Fig. 3.22. LEDs at 905 nm are illuminated in the reference modules, and this light propagates through the precision 0.2 mm apertures in the modules, through the half lenses, where they are collimated, and then into the alignment telescope. Since the telescope is collimated, when the images of the apertures are in focus at the focal plane of the alignment telescope, they will also be in focus in the optical assembly. A CCD camera sensitive

Table 3.8 Steps in the optical assembly alignment procedure

1.	Collimate alignment telescope and attach CCD camera (no IR filter)
2.	Attach reference modules to the optical assembly
3.	Fix optical assembly to the telescope stand, parallel to the telescope
4.	Power up 905 nm LEDs in the reference modules
5.	Adjust height of optical base plate to focus module aperture images
6.	Fix positon of optical base plate
7.	Adjust X-Y position of modules until images are centred and coincident
8.	Fix position of X-Y adjust plates for each module
9.	Iterate steps 7 and 8 if slight adjustment is needed
10.	Secure optical base plate and X-Y adjust plates
11.	Remove reference modules and attach laser and detector modules

Fig. 3.22 *Left* Collimator in its mounting assembly (note that the telescope mirror objective is at ceiling height). *Right* Optical assembly in position under collimator

to near infrared light is placed at the focal plane of the telescope and used to view the images of the apertures. When the circular image from each channel is focused, centred, and coincident with the other, optical alignment has been achieved. The adjustable mechanics of the system can then be secured and the reference modules removed and replaced by the pre-aligned laser and detector modules.

3.12 Thermal Effects on Optical Performance

Since the prototype instrument needs to be able to perform over a wide range of temperatures, some analysis of possible effects of this temperature range was required. The two key thermal considerations are focus changes with temperature, and changes in the metal-adhesive-glass system in the lens mount plate.

3.12.1 Focus Displacement Effects on Signal to Noise Ratio

An important consideration regarding the opto-mechanical design is the influence temperature changes have on the optical performance of the instrument. Maintaining adequate focus is of primary concern, as both the lenses and the mechanical assembly change with temperature. Fortunately, focal length changes of the lenses due to temperature changes can be calculated in a straightforward manner using ray-tracing software, and, since the optical tube assembly is nearly 100 % aluminium, the focus shifts due to changes in the length of the tube assembly can be accounted for using the linear thermal expansion coefficient of aluminium, 23.1×10^{-6} m/m K [22].

For the custom aspheric lenses that have been designed, an OSLO model of the lens was used to calculate that the focal length decreases by 7 μm/°C increase in temperature. The aluminium mechanical assembly holding the optics, on the other hand, increases in length with temperature. A 335 mm length of aluminium increases by 7.7 μm/°C. Since the focal length decreases with temperature and the mechanical assembly length increases with temperature, the optical and mechanical shifts do not cancel each other out but are instead additive; the plane in which the detector and laser are located is moved 14.7 μm down from the focal plane for each degree increase in temperature. Therefore, the effect of temperature change on optical signal to noise ratio must be considered.

In order to determine the size of the image of the laser at various defocus points, the image of the laser stripe as seen by the detector is determined by the same convolution method used in Sect. 3.9.1. But here, instead of being convolved with the minimum resolvable spot size of the lens at focus, the laser stripe image was convolved with the larger spot size at a defocused point. The following assumptions were made in this calculation. First, the power distribution of the laser is uniform and is spatially incoherent. Secondly, the spots formed by each half-lens (cut from the same whole lens) are equal to each other in size at each defocus point. Thirdly, the laser and the detector always defocus by the same amount. In addition, this calculation is used to compare SNRs only for ranges beyond full overlap. Note that in contrast to Sect. 3.9.1, here it is the calculated, not the measured spot size that is used in the convolution.

As an example, at 0.9 mm defocus, the fraction of energy in the defocused image that is collected by a detector with a radius of 250 μm was determined as shown in

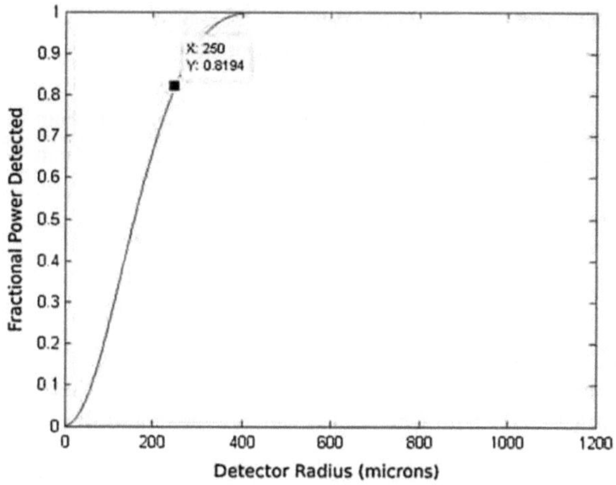

Fig. 3.23 Fractional laser power collected by 0.25 mm radius detector with +0.9 mm defocus of each channel

Fig. 3.23. Here it is shown that 82 % of the received light is incident on the detector when the system is defocused by 0.9 mm.

The same calculation was performed for a number of defocus distances as shown in Table 3.9. Here for each defocus in the first column, the calculated minimum spot radius is given in the second column and the fractional power collected by a 0.25 mm radius detector at this displacement is given in the third column. It is shown that up to a defocus of 0.72 mm, greater than 90 % of the collected laser energy should be seen by the detector.

To investigate the SNR implications of temperature changes over the required operating temperature range of the instrument (−40 to +60 °C), calculations were done considering both the effective focal length changes of the lenses and the changes

Table 3.9 Effect of defocus on collected laser power

Laser and detector defocus (mm)	Minimum spot radius (mm)	Fractional power on r − 0.25 mm detector
0.55	0.14	0.99
0.6	0.15	0.98
0.7	0.17	0.94
0.72	0.18	0.92
0.82	0.20	0.87
0.9	0.22	0.82
1.0	0.25	0.74
1.25	0.31	0.60
1.5	0.37	0.49

Table 3.10 Effect of temperature shift on signal to noise ratio

Assembly material	Defocus distance, SNR factor for Δ60 °C	Defocus distance, SNR factor for Δ40 °C
Aluminium	0.9 mm → 0.82 SNR	0.6 mm → 0.98 SNR
Aluminium Inc. 0.1 mm focus tol.	1.0 mm → 0.74 SNR	0.7 mm → 0.94 SNR
Steel	0.7 mm → 0.94 SNR	0.4 mm →>0.99 SNR
Steel Inc. 0.1 mm focus tol.	0.8 mm → 0.87 SNR	0.5 mm →>0.99 SNR

in a 335 mm tube length aligned at 20 °C, for shifts of −60 and +40 °C, for two different materials, aluminium and steel. The coefficient of thermal expansion for structural steel is 12×10^{-6} m/m K [22], approximately half that of aluminium. The possible extensions of the focal length due to the 0.1 mm focus tolerances were also considered for all of these cases. The results of this analysis are shown in Table 3.10. Note that the SNR effects for cooling or heating by the same amount were assumed to be the same by geometric optics.

What this means is that even though steel undergoes half the length shift that aluminium undergoes for a given temperature change, the length changes of aluminium over the operating temperature range are not extreme enough to limit the performance significantly. Since aluminium is considerably lighter and easier to machine than steel, it is preferred as the primary material used for mounting of the optical assembly. In order to limit the effects of extreme cold, a heater is used to keep the optical assembly from getting colder than −20 °C. This means that if the optics are aligned at +20 °C, the maximum temperature change it should undergo in either direction is 40 °C, which in turn means that even with the introduction of focusing tolerances of ±0.1 mm, the SNR of the instrument should not drop below 90 % of its optimum value over the operating temperature range of the instrument.

3.12.2 *Effects of Thermal Changes to the Lens Mount Sub-assembly*

Mounting two half-lenses in the same plate is somewhat more complex than mounting a circular lens directly into a tube assembly. The lenses were designed with a flat base to help ease the mounting process, but there are a few other important considerations. The lens mount plate is shown in Fig. 3.24.

The first issue to consider is thermal displacement of the optical axes. The half-lenses are mounted such that the flat edge of the cut diameter of each plate is intended to be kept tight along the edge of the lens plate it faces. If this is maintained on both sides, the optical axes can be expected to be pushed apart or pulled together slightly as the aluminium plate expands or contracts, respectively. Since there is a 21 mm offset between the optical axes at 20 °C, this can be expected to change by +/−0.019 mm

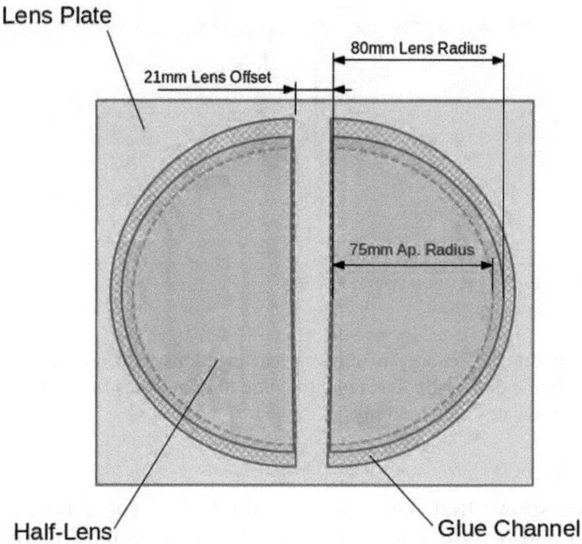

Fig. 3.24 Schematic of lens mount sub-assembly. Note that the clear aperture radius of the lens plate is 75 mm

over the maximum temperature change of +/−40 °C. As the laser and detector are both mounted in aluminium with the same initial gap between them, they can be expected to experience parallel, matched displacements to the optical axes, thus maintaining straight alignment of both channels.

The second thermal issue for this sub-assembly is the mechanical behaviour of the adhesive over the operating temperature range of the instrument. The gap in the lens plate around the semi-circular side of each half lens allows for differences in thermal expansion of the aluminium and glass, and also allows a channel for the application of adhesive. Due to the large size of the lenses, rigid adhesives are undesirable as even small thermal expansion mismatches can lead to large stresses on the glass which can adversely effect the optical quality of the lenses, or worse, crack them. Therefore, a flexible, high quality silicone sealant is applied in this channel around the half-lens and a fine line of it placed on top of the flat edge of the metal facing the cut side of the lens. This method has proven to be successful in terms of mitigating stresses, but it has been found on occasion to introduce slight tilts of up to 1° on each half-lens. While it is expected that a mechanical solution to this problem will be found prior to production of the instrument, the effect of this tilt on the performance of the lenses was characterised using ray tracing in OSLO. The lens was tilted by an angle of 2° in order to simulate the maximum expected distortion of the signal due to two trips through lenses tilted at angles of 1° each. The fraction of radial energy contained within a given radius was calculated using the point spread function at this angle as shown in Fig. 3.25.

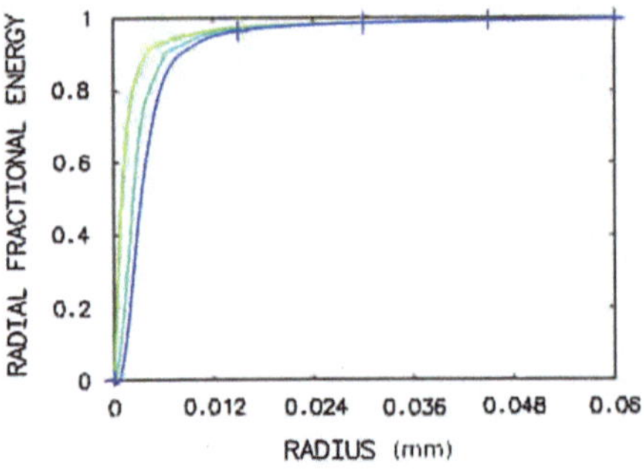

Fig. 3.25 Radial fraction of energy of custom aspheric lens with no tilt (*left*) and with 2° tilt (*right*). The *green line* indicates on axis rays, the *blue line* indicates full-field (1.5 mrad) rays, and the *blue-green line* indicates half-field rays

This analysis shows that if one lens is tilted at 2° (or if both lenses are tilted at 1°), the light can be collected within a radius of 0.15 mm. Applying this as the minimum spot radius, Table 3.9, shown previously, can be used to determine that 98 % of the imaged light would fall within the 0.25 detector radius at 20 °C where no temperature-induced defocus is expected. However, it is presumed that this tilt effect would not generally occur at the alignment temperature of 20 °C, but rather at extreme temperatures where temperature-induced defocus effects also occur. If the minimum spot radii from these effects are summed in order to approximate their combined effects, at a temperature increase of 40 °C, the aluminium-design instrument would see a minimum spot radius of 0.17 mm (spot radius due to defocus) +0.15 mm (spot radius due to tilt), or 0.32 mm. By looking up this spot radius in Table 3.9 it can be seen that the fractional power received on the detector could be as low as 60 % of the imaged power and hence 60 % of the optimum SNR. However, this can be deemed acceptable considering that the estimated SNR advantage of this design over instruments of a similar class is 4.3 and therefore the minimum SNR after 40 % reduction should still be 2.6 times greater than the reference. In addition, since these calculations were done, a number of test gluings of half-lenses showed maximum tilts on the order of 0.1° at extreme temperatures. Therefore the lens tilt effect is not considered to have significant influence on the signal to noise ratio.

3.13 Conclusion

A prototype lidar ceilometer based on a novel divided-lens optical configuration was designed and built in a full environmental enclosure. A photo of the built-up prototype is shown in Fig. 3.26. The optimisation of the optical design for good signal to noise ratio and close-range overlap has been discussed, along with manufacturing considerations such as use of low cost components and ease of assembly and

Fig. 3.26 Prototype instrument in full external enclosure with cowelling

alignment. Characterisation processes for components were used to investigate the influence of various parameters on the optical signal to noise ratio of the instrument, and the trade-offs involved in the design process have been emphasised. Although the final instrument refinements and costing are still in progress, the design has already proven to be an excellent compromise between performance and cost. Compared to leading instruments of similar performance class, the maximum range of this design is approximately 25 % greater and the estimated selling price about 25 % lower.

References

1. Federal Aviation Administration (FAA), Automated weather observing systems (AWOS) for non-federal applications, advisory circular 150/5220-16D, FAA, United States Department of Transportation, Technical Report 2011
2. BSI, BS EN 60825–1:2007, Safety of laser products-Part 1: Equipment classification and requirements, BSI Std. (2007)
3. J. Kallio, U.S. Patent 7,428,041, 23 Sept 2008
4. E. Gregorio, Design methodology of a ceilometer lidar prototype, in *Proceedings of the IEEE International Geoscience and Remote Sensing Symposium* (2007), pp. 3162–3165

5. E. Gregorio, F. Rocadenbosch, J. Tiana-Alsina, A. Comern, R. Sanz, J. R. Rosell-Polo, Parameter design of a biaxial lidar ceilometer. J. Appl. Remote Sens. **6**(1), 063 546-1-063 546–19 (2012)
6. G. Gimmestad, D. Roberts, J. Stewart, *L.I.D.A.R. Engineering Short Course Notes*. Georgia Institute of Technology, Atlanta (2007)
7. C. Münkel and J. Räsänen, in *New Optical Concept for Commercial Lidar Ceilometers Scanning the Boundary Layer*, ed. by A. Comeron, M.R. Carleer, R.H. Picard, N.I. Sifakis, SPIE, vol. 5571, issue(1), pp. 364–374
8. J. He, W. Liu, Y. Zhang, R. Kan, Z. Chen, J. Ruan, Atomosphere boundary layer height determination and observation from ceilometer measurements over Hefei during the total solar on July 22, eclipse. Chin. Opt. Lett. **8**(5), 439–442 (2009)
9. G. Martucci, C. Milroy, and C. D. ODowd, Detection of cloud-base height using Jenoptik CHM15k and Vaisala CL31 ceilometers. J. Atmos. Oceanic Technol. **27**(2), 305–318 (2010)
10. R. N. Clark, *Manual of Remote Sensing, Volume 3, Remote Sensing for the Earth Sciences*. Chapter 1: Spectroscopy of Rocks and Minerals, and Principles of Spectroscopy (Wiley, New York, 1999), pp. 3–58
11. Oriel Instruments. Introduction to solar radiation. Oriel instruments, a Newport Corporation Brand (2012), http://www.newport.com/Introduction-to-Solar-Radiation/411919/1033/content.aspx
12. Y. Saito, H. Kurata, H. Kurushima, F. Kobayashi, T. Kawahara, A. Nomura, T. Maruyama, M. Tanaka, Experimental discussion on eye-safe 1.54 m photon counting lidar using avalanche photodiode. Opt. Rev. **11**, 378–384 (2004). doi:10.1007/s10043-004-0378-7
13. L. Matthies, P. Bellutta, and M. Mchenry, Detecting water hazards for autonomous off-road navigation, in *Proceedings of SPIE Conference 5083: Unmanned Ground Vehicle Technology V* (2003), pp. 263–352
14. L. Sauvage, S. Lolli, B. Guinot, and M. Lardier, New eye safe autonomous EZ lidar for pollution and meteorological continuous monitoring, in *World Meteorological Organization (WMO) Technical Conference on Meterological and Environmental Instruments and Methods of Observation* (2008)
15. P. Rainbow, "3JMTTFestimate", *Laser Components* (Unpublished, GmbH, August 2005)
16. R. Paschotta, *Avalanche Photodiodes. RP Photonics Encyclopedia of Laser Physics and Technology* (RP Photonics Consulting, GmbH, 2012). http://www.rp-photonics.com/avalanche_photodiodes.html
17. Andover Corporation (2012), http://www.andovercorp.com/Web_store/General_info/Technical.php
18. American Society for Testing and Materials (ASTM). Terrestrial reference spectra for photovoltaic performance, http://rredc.nrel.gov/solar/spectra/am1.5/ASTMG173/ASTMG173.html
19. Lambda Research. OSLO: Optics Software for Layout and Optimization (2012). http://lambdares.com/oslo (August, 2014)
20. Taylor Hobson. Optics PGI surface profilers (2012), http://www.taylor-hobson.com/products/10/64.html#PGI-Dimension2 (August, 2014)
21. The Mathworks, Inc. MATLAB: The language of technical computing (2012), http://www.mathworks.co.uk/products/matlab/
22. G. Elert, Thermal expansion. The Physics Hypertextbook (2012), http://physics.info/expansion/

Chapter 4
Determination of Lidar Overlap

4.1 Introduction

A good understanding of overlap of transmitter and receiver channels as a function of height is essential for interpretation of lidar returns. At the very least, the range at which full overlap is first reached should be known; this can be used to set a minimum range from which signal inversion can be performed. If information prior to full overlap distance is required, the overlap profile must be known accurately. If the overlap function is not accurately known then estimates of the aerosol backscatter coefficient at low altitudes will be poor [1]. Because of this, lidar systems frequently invert the lidar equation using only data collected beyond the point where there is high confidence that full overlap has been reached. If overlap can be predicted, however, aerosol extinction in the lowest region of the atmosphere can be measured more accurately, as can derived parameters such as vertical visibility.

As discussed in Chap. 2, for the case of weak, elastic scattering from atmospheric constituents, the backscattered intensity is related to transmission and the backscatter coefficient by the lidar equation. Under the assumption that the effects of multiple scattering are negligible, the lidar equation in its simplest form expresses the backscattered signal power, $P(r)$, at a given wavelength as a function of target range, r, such that [2]

$$P(r) = K_s G(r) \beta(r) T(r), \tag{4.1}$$

where K_s is the system factor (which includes the area A_0 of the receiver), $\beta(R)$ is the backscatter coefficient at distance r, and $T(r)$ is the transmission term that describes round-trip losses as the laser pulse travels through the atmosphere to and from range r. The remaining factor, $G(r)$, the geometric factor, is the concern of this chapter and is generally stated as

$$G(r) = \frac{O(r)}{r^2}, \tag{4.2}$$

J.D. Vande Hey, *A Novel Lidar Ceilometer*, Springer Theses,
DOI 10.1007/978-3-319-12613-5_4

where $O(R)$ is the overlap function describing the fraction of laser beam cross-section imaged by the receiver as a function of range. In this expression, the quadratic factor in the denominator is due to the reduction in solid angle subtended by the lidar receiver and is consequently inversely proportional to the square of the distance from the target to the entrance pupil [2]. When a lidar is in normal use it is noted that this distance is simply equal to the range, however, if additional optics are included between the lidar and the target, the position of the entrance pupil changes and for this reason in this chapter it is more useful to write the geometrical factor in the explicit form,

$$G(r) = \frac{O(r)}{{r_p}^2}, \tag{4.3}$$

where r_p is the distance from the target to the entrance pupil of the instrument.

For situations in which the field of view of the receiver is greater than the divergence angle of the laser, Stelmaszczyck et al. [3] formulated a precise definition for the overlap function, or geometrical compression form factor, as the ratio of the energy E_{det} collected by the photodetector to the energy E_{obj} incident on the primary objective of the receiver such that

$$O(r) = E_{\text{det}}(r)/E_{\text{obj}}(r) \tag{4.4}$$

Stelmaszczyk et al. noted that once the overlap function has been determined, it can be used to define the telescope's effective receiver area, A_{eff}, such that

$$A_{\text{eff}}(r) = O(r)A_0, \tag{4.5}$$

where A_0 is the area of the receiver's primary objective [4].

When $O(r)$ is small, much of the collected light is focused outside the detector. These losses tend to decrease with distance, increasing the overlap until all backscattered radiation incident on the lens is registered and $O(r) = 1$. The closest range at which this occurs is called the full-overlap distance. At this point and beyond, the laser beam stays within the receiver FOV and each point on the primary lens or mirror of the receiver has the same light collecting efficiency. This doesn't hold true if the laser beam diverges faster than the receiver FOV, or if the laser is tilted so much that it exits the receiver FOV at some range.

Note that here $O(r)$ refers to the overlap function (where $O(r) = 1$ when the laser beam cross section is fully imaged by the receiver onto the detector). It does not refer to the fractional overlap area, $\mathcal{A}(R)/\pi(w(R))^2$. The overlap area $\mathcal{A}(R)$ is the area of the region at range R where the receiver field of view overlaps the laser beam of radius $w(R)$ as shown in Fig. 4.1. It is possible to make the transmitter beam and receiver field of view coincident and therefore make the fractional overlap area identically 1 by using a coaxial arrangement with common optics such that the laser and detector use exactly the same aperture. However, as stated by Measures [4], it is not possible to make the overlap function

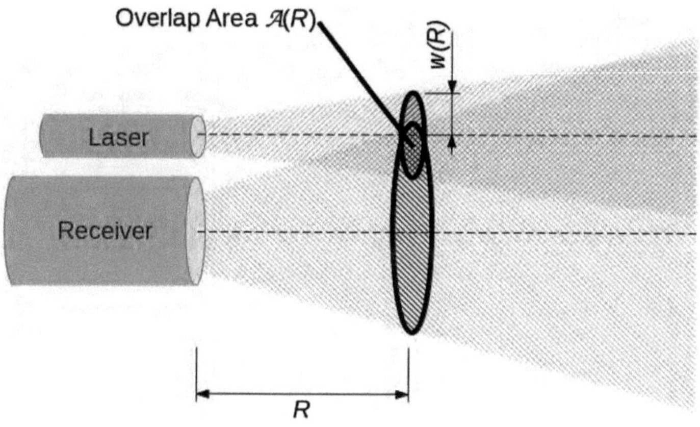

Fig. 4.1 Overlap area of a biaxial lidar system at range R

$O(r) = 1$ throughout the range of the instrument unless the aperture of the receiver objective is the only aperture in the system. Detector apertures used in lidar systems are usually, if not always, set back from and much smaller than the receiver objective aperture. Since scattering from atmospheric particles occurs over a wide range of angles, light backscattered from close ranges will focus over a broad region as shown in Fig. 4.2. Here it is shown that when a scatterer at approximately two focal lengths from the lens is illuminated by the laser beam (shown in light red/solid grey in print), only a small fraction of the backscattered light incident on the lens (shown in grey hatching) is focused onto the detector. Whether lidar systems use biaxial optics [5], expand the laser beam outside of the receiver optics [6] or utilise separate regions of the aperture for the laser and receiver [7], overlap correction is required for calibration of measurements at near ranges.

This chapter describes the work that has been undertaken in order to evaluate overlap in the prototype instrument. Following a review of literature discussing overlap determination, a convolution method that was devised in order to account for the unique geometry of the prototype optical system is described. This geometric calculation technique, which was validated by experiment and through comparison with other geometric methods, is the tool that was used to evaluate the various possible ceilometer designs discussed in Chap. 3. After the calculation is explained, a novel optical test system capable of measuring the overlap of a lidar system is proposed. The optical test system can be considered as a means of presenting a "virtual cloud" that appears at a prescribed distance from the lidar instrument; the virtual cloud can be positioned at various effective ranges and the returns from each range measured in order to determine the overlap. It provides a significant advantage over existing methods for elastic lidar overlap determination because it can be performed in the space of a laboratory environment with controlled optical conditions. Finally, the results of the calculation and measurement are compared with horizontal lidar returns from a hard target translated along the overlap region.

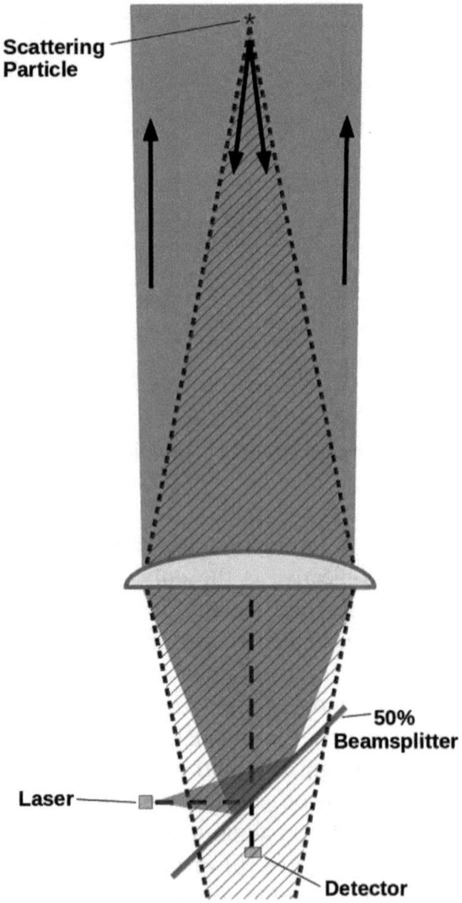

Fig. 4.2 Focusing of light received from close ranges

4.2 Approaches to Overlap Determination

In this section, theoretical techniques for determination of overlap described in the literature are summarised. These methods are based almost entirely on geometric optics. After this, experimental techniques from the literature relevant to elastic lidar are described.

4.2.1 Review of Overlap Calculation Techniques in the Literature

Theoretical determination of the overlap function has been discussed by several authors. Halldórsson and Langerholc employed a method [8] whereby a geometric optics approach is applied by integrating over scattering angles to determine the fraction of light focused by the telescope onto the detector from each infinitesimal scattering volume. Their work included an analytical determination of the cross-sectional area of the overlap region in order to determine the illuminated region at each distance that overlaps with the receiver field of view. The primary drawback of this method is its reliance on explicit determination of the overlap area in order to define the region over which to integrate, which adds unnecessary complexity to the calculation.

Harms et al. [9] presented a method that used integration in object space, that is, in the sensing volume, over a region bounded by the area of the image of the detector. They expressed the effects of overlap in terms of what they termed a geometric compression of the signal, and calculated the irradiance and power incident on detectors of different sizes in the focal plane using fixed extinction and backscatter for a clear atmosphere. In order to accomplish this, however, their derivation included the use of lidar equation variables, such as laser pulse length and atmospheric attenuation, that do not directly affect the overlap. Ancellet et al. [10] went on to refine this method by including a thorough treatment of laser distribution. As with the previous method, however, the method of Harms et al. includes unnecessary complexity, here in terms of extra variables, if only determination of the overlap function is desired.

Kuze et al. [11] developed a method by which rays are projected by geometrical optics in order to formulate analytical expressions for ranges at which overlap onset and full overlap occur. The overlap function is derived from the analytical expressions for these heights by an arbitrary function (a hyperbolic tangent function) along with a fitting parameter. The fitting parameter can then be adjusted to match profile shapes from observed system overlap to improve the accuracy of the analytical expressions. This method provides a straightforward means of expressing overlap of parallel biaxial lidars and was shown to generate simulated lidar returns for specified atmospheric conditions that matched observed returns for a number of different receiver field of view angles reasonably well. The advantage of this method is its flexibility for refinement through observation via the fitting parameter. However, in the absence of well-understood atmospheric conditions in which to verify the overlap calculation through experiment, a method relying only on the optical parameters of the instrument is preferred.

Stelmaszczyk et al. [3] followed an approach similar to that of Kuze et al. to develop an analytical expression for overlap, but extended it to account for laser tilt and coaxial lidar configurations. In addition, they removed the need for an arbitrary fitting function. Their method used an angular approach to determining the overlap of the detector aperture with the image of the laser beam cross section in the focal plane as a function of height for a uniform laser distribution. The concept of this method, that of considering images from the perspective of the detector, is similar to the one presented in the next section, but the mechanics of its derivation are different in that it formulates the problem in terms of angular image projection. While the method used by Stelmaszczyk et al. assumes a uniform distribution and is used with systems of circular objective, the approach applied in this chapter allows for variation of the laser power distribution and objective aperture shape, and also considers the particular case of a diode laser of short coherence length.

A few additional approaches have been discussed in the literature. Sassen and Dodd [12] developed an approximation for determining overlap area as a function of range using a Gaussian laser distribution, but this method did not account for close range imaging effects. Velotta et al. [13] applied a ray-tracing approach, and Kokkalis et al. [14] extended the ray-tracing approach to include effects of additional components, such as the optical bandpass filter, on the overlap. Ray-tracing utilises the same geometrical optical principles used by any of these calculation methods and should therefore arrive at the same solution. It allows optical component models to be used, which can improve the accuracy of the method, but it does not provide a mathematical expression for overlap.

Finally, a hybrid approach using geometrical optics along with diffraction theory has been discussed by Berezhnyy [15]. Diffraction is an important consideration that should not be ignored in a real optical system. However, since Berezhnyy found that diffraction effects were small for receivers without central obstructions, and since the objective aperture size is very large compared to the wavelength, the additional complexity required for considering diffraction effects was not deemed to be essential for the analysis here.

4.2.2 Review of Overlap Measurement Techniques in the Literature

A number of experimental techniques applicable to elastic backscatter lidar have utilised atmospheric measurements to determine overlap. The most straightforward of these were described by Sasano et al. [1], who used measurements of a well-mixed boundary layer, and Tomine et al. [16], who used measurements of a light mist to determine the overlap. Both of these approaches compute an overlap function that modifies a measured signal to match the exponentially decreasing return expected from a homogeneous scattering volume. Given truly homogeneous atmospheric conditions throughout the overlap region, this approach is valid provided the detector response is linear throughout dynamic range of the signal. However, in the case of the measurements by Sasano et al., while a well-mixed boundary layer with a nearly uniform distribution of aerosols may occur regularly, at the time of measurement it may not be uniform throughout the overlap region, particularly for lidar systems with narrow fields of view and therefore large overlap distances. In addition, an eye-safe lidar instrument such as a ceilometer may require a large averaging time in order to get a good aerosol profile due to limited laser output, and during the measurement time required the atmospheric conditions may change. The returns from a mist would be considerably larger than those from aerosols, making the approach of Tomine et al. more desirable for use with instruments of relatively low signal to noise ratios. However, in many locations occurrence of conditions of light mist may be rare and difficult to predict. In addition, even with the regular occurrence of light mist, the assumption of homogeneity of droplet size and concentration throughout the overlap region would need to be validated somehow as precipitation intensity is often inconsistent.

Another experimental approach was suggested by Dho et al. [17]. They applied a fifth-order polynomial regression technique to data obtained beyond the height of full overlap to extrapolate the atmospheric conditions below this point. By comparing the extrapolated atmospheric conditions to the return signal below the full overlap point, the overlap function was established. While the results shown were reasonable for one particular measurement, the assumption that this technique could be extended to any heterogeneous atmospheric conditions is questionable since, for example, the likelihood of heterogeneous aerosol concentrations higher in the atmosphere being linked to aerosol concentrations lower down by a fifth-order polynomial that describes the whole distribution seems highly unlikely.

Techniques for measurement of the overlap function of an elastic plus Raman lidar were developed by Wandinger and Ansmann [18] and by Hu et al. [19]. Raman lidars exploit the Raman scattering effect [20] whereby the incident wavelength is shifted by predictable changes to the vibrational-rotational energy levels of the scattering molecules. Raman lidar channels are typically used to measure returns from molecular nitrogen or oxygen, whose concentrations in the atmosphere are well understood and whose backscatter coefficients and transmissions are known based on their concentrations. Under the assumption that the overlap function for the elastic channel is identical to that of the Raman channel, which Wandinger and Ansmann state is reasonable for a well-aligned system based on experience, the overlap function for both channels can be found by looking at what percentage of the expected molecular returns are received by the Raman channel at each range. This method, however, is not applicable to elastic-only lidars.

Another experimental approach, described by Guerrero-Rascado et al. [21], utilised returns from ceilometers which were assumed to have well-understood, small full-overlap ranges in order to use the signals beyond this range to calibrate the overlap of another elastic lidar system. If an instrument of known overlap is available, this method is feasible. However, errors in the judgement of full overlap height and/or non-linearity in the receiver response of the reference instrument can introduce considerable error into the overlap calibration of the other instrument.

The experimental method described later in this chapter allows the overlap of an elastic lidar system to be measured with no assumptions about atmospheric conditions except for those within a small laboratory that contains the setup. Provided that optical instrumentation used for the measurement is well characterised, this approach can be used successfully for overlap determination.

4.3 Geometric Overlap Calculation

In order to be able to relate experimental determination of the prototype's overlap to theory, a method for calculation of overlap based on geometric optics was devised. This section describes that method, compares it with methods from the literature for simpler optical arrangements, and discusses a few examples of overlap calculated for different optical designs.

The problem of geometric determination of lidar overlap can be approached by considering fractional power incident on the detector using a calculation method here called the *convolution method*. The only information needed for this calculation is the angular laser power distribution and the geometric arrangement of the optical system, including the laser emitter pattern, lens focal lengths, offset of optical axes, and aperture profiles. As described previously, the prototype lidar that has been developed is a biaxial system utilising aspheric lenses of focal length $f = 335$ mm (at 905 nm), a pulsed diode laser operating at 905 nm and a silicon APD detector. One small but important difference between the final prototype design and the system considered in this chapter is that a single-stack 70 W laser was used here rather than the double-stack 135 W laser that was eventually selected for the design. The double-stack laser system, the overlap of which was shown in Fig. 3.7, was calculated to have a slightly longer full overlap distance than that of the system considered here. Note that while the calculation is derived here specifically for the unique divided lens geometry and laser diode used in the instrument that was evaluated, it can be adjusted for a wide variety of optical configurations.

The essence of the *convolution method* is as follows. At each measurement range, each infinitesimal element of the laser emitter can be considered to illuminate a region whose area is defined as a projection of the transmitter objective aperture. The total laser power distribution at each range is the collective power of all of the laser element projections. The distribution at each range is found by convolving the appearance of the transmitter objective as seen from that range with the appearance of the image of the laser as seen from that range. At close ranges, this convolved distribution looks like the transmitter aperture with slightly extended blurred edges, and at far ranges it looks like the image of the laser emitter with slightly extended blurred edges. Between the near and far range there is a gradual transition from one of these distributions to the other. Similarly, each point on the detector is considered to be sensitive to light backscattered to the instrument from the region contained within the projected area of the receiver objective aperture at each given range. The effective receiver sensitivity function is calculated in the plane at each range as the convolution of the receiver objective as seen from that range with the appearance of the detector aperture as seen from that range. Here again, at close ranges the convolved distribution looks like the receiver objective, at far ranges it looks like the image of the detector aperture, and in between it varies gradually. If the laser power distribution is normalised, the overlap function can be found as the product of the laser distribution and the effective receiver sensitivity function at each range.

In order to explain this method explicitly, a general function for the laser distribution is formulated first, along with its normalisation. Then the expression describing the effective received intensity is derived. Finally, the laser power function is expressed fully and the full analytical expression for overlap is given.

4.3.1 Convolution Method Overlap Calculation

The derivation of this calculation is most straightforward when approached first from the standpoint of the receiver. It is simplified somewhat, however, if a general function describing the transmitted laser power in the plane at range r, along with this function's normalisation are given first. Suppose that the cross-sectional laser intensity incident on a uniform scattering plane at some distance, r, from the instrument is given by the function $\Upsilon(x_T, y_T, r)$. As the laser beam propagates, it diverges, but in the absence of attenuating particles (i.e., in a vacuum) the total laser power remains constant and integrating over a region containing a full cross section of the beam at any height gives the same power. Therefore the total power illuminating the scattering plane at any height can be normalised by introducing the constant Υ_0 such that

$$\Upsilon_0 \int_{-\infty}^{\infty} \int_{-\infty}^{\infty} \Upsilon(x_T, y_T, r)\, dx_T\, dy_T = 1. \tag{4.6}$$

This normalised general function for the laser power distribution is sufficient for the derivation of an expression for fractional received power. The specifics of the transmitter function are discussed later in this section.

If the overlap as a function of range is defined as the fraction of light incident on the receiver objective that is directed onto the detector, it can be expressed as

$$O(r) = \frac{P(r)}{P_{D=\infty}(r)}. \tag{4.7}$$

Here $P(r)$ is the backscattered power incident on the actual detector area, and $P_{D=\infty}(r)$ is the power that would be incident on a detector of infinite aperture, therefore equal to the amount of backscattered light incident on the receiver objective.

In order to calculate the power $P(r)$ collected from various ranges, the receiver geometry must be considered. In Fig. 4.3 it is shown that geometrically, each point on the detector in the focal plane collects light from scattering particles within a column whose cross section is equal to that of the projected lens aperture. In the object plane (the plane of measurement at range r) this column is displaced from the optical axis by a distance determined by the magnification. At any range r the receiver field of view includes all projected apertures. Since the detector apertures and therefore the projected angles involved are small, it is assumed that each of the projections of the lens aperture can be considered to have the same cross-sectional area as the lens aperture itself and lie in the plane at range r. Note that in reality, however, slight curvature of the surface formed by the projections at a specific distance with varying angle would result, and slight variation in the effective shape of the lens aperture as viewed from different points on the detector would occur.

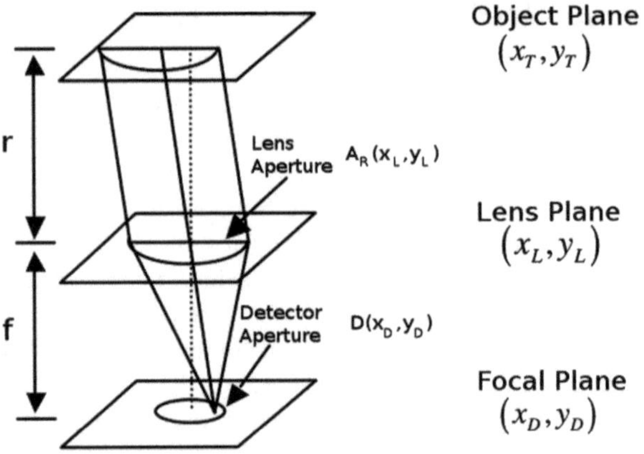

Fig. 4.3 Geometry of the receiver

The receiver half lens aperture can be defined as an indicator function $A_R(x_L, y_L)$ in the lens plane such that

$$A_R(x_L, y_L) = \begin{cases} 1 & (x_L^2 + y_L^2) < (\frac{d_L}{2})^2, \quad x_L < 0 \\ 0 & \text{otherwise,} \end{cases} \tag{4.8}$$

where d_L is the clear aperture diameter of the lens from which the half lens was cut. The receiver's optical axis, in the lens plane, lies at $(x_L, y_L) = (0, 0)$.

Since the intensity $I(x_D, y_D, r)$ from range r incident on a point (x_D, y_D) in the focal plane is proportional to the integral of the laser intensity $\Upsilon(x_T, y_T, r)$ within the corresponding projection of the lens aperture $A_R(x_L, y_L)$ on the uniform scattering surface in the object plane at distance r from the instrument, the intensity at this point on the detector can be expressed

$$I(x_D, y_D, r) = \epsilon(r)\Upsilon_0 \int_{-\infty}^{\infty}\int_{-\infty}^{\infty} A_R(\frac{r}{f}x_D + x_T, \frac{r}{f}y_D + y_T)\Upsilon(x_T, y_T, r)\, dx_T\, dy_T, \tag{4.9}$$

where $\frac{r}{f}$ is the magnification determined from the range r and the focal length f, and $\epsilon(r)$ is a function that relates intensity in the scattering plane to intensity incident on the receiver objective. Here the form of $I(x_D, y_D, r)$ is a convolution with a kernel defined by the aperture of the receiver.

Now that the intensity has been found for each point on the detector, the total power received across the area of the detector can be found. First, the circular detector aperture of diameter d_D in the focal plane is defined by the indicator function $D(x_D, y_D)$ such that

$$D(x_D, y_D) = \begin{cases} 1 & (x_D^2 + y_D^2) < (\frac{d_D}{2})^2 \\ 0 & \text{otherwise.} \end{cases} \tag{4.10}$$

The total power $P(r)$ from range r incident on the detector can then be found by integrating the intensity in the focal plane across the region bounded by the detector such that

$$P(r) = \int_{-\infty}^{\infty}\int_{-\infty}^{\infty} I(x_D, y_D, r) D(x_D, y_D)\, dx_D\, dy_D. \tag{4.11}$$

Note that introducing the function $D(x_D, y_D)$ makes it possible for the integral in Eq. 4.11 to be written with infinite limits of integration.

When $I(x_D, y_D, r)$ is substituted from Eq. 4.9, $P(r)$ becomes

$$P(r) = \epsilon(r)\Upsilon_0 \int_{-\infty}^{\infty}\int_{-\infty}^{\infty} \Big[\int_{-\infty}^{\infty}\int_{-\infty}^{\infty} A_R(\frac{r}{f}x_D + x_T, \frac{r}{f}y_D + y_T) \cdot \Upsilon(x_T, y_T, r) D(x_D, y_D)\, dx_T\, dy_T \Big] dx_D\, dy_D. \tag{4.12}$$

Under the substitution of variables $u = \frac{r}{f}x_D$ and $v = \frac{r}{f}y_D$, this can be rewritten such that

$$P(r) = \epsilon(r)\Upsilon_0 \int_{-\infty}^{\infty}\int_{-\infty}^{\infty} \Big[\int_{-\infty}^{\infty}\int_{-\infty}^{\infty} A_R(u + x_T, v + y_T) \cdot \Upsilon(x_T, y_T, r) D(\frac{u}{(\frac{r}{f})}, \frac{v}{(\frac{r}{f})}) dx_T\, dy_T \Big] \frac{du\, dv}{(\frac{r}{f})^2}. \tag{4.13}$$

Because the functions A_R and Υ have compact support and are bounded, the order of integration may be reversed to give

$$P(r) = \epsilon(r)\Upsilon_0 \int_{-\infty}^{\infty}\int_{-\infty}^{\infty} \Upsilon(x_T, y_T, r) \Big[\int_{-\infty}^{\infty}\int_{-\infty}^{\infty} A_R(u + x_T, v + y_T) \cdot D(\frac{u}{(\frac{r}{f})}, \frac{v}{(\frac{r}{f})}) \frac{du\, dv}{(\frac{r}{f})^2} \Big] dx_T\, dy_T. \tag{4.14}$$

The expression in square brackets in Eq. 4.14 may be interpreted as the weighting function that determines the fractional effective receiver area at range r. It describes the image of the detector in the object plane as the sum of all projections of the receiver aperture.

If an infinite detector were used, the detector function would become the constant function

$$D\left(\frac{u}{(\frac{r}{f})}\right) = 1. \tag{4.15}$$

This means that all of the light from the scattering plane incident on the receiver lens would be detected, and Eq. 4.14 could be expressed

$$P_{D=\infty}(r) = \epsilon(r)\Upsilon_0 \int_{-\infty}^{\infty}\int_{-\infty}^{\infty} \Upsilon(x_T, y_T, r) \times \left[\int_{-\infty}^{\infty}\int_{-\infty}^{\infty} A_R(u + x_T, v + y_T)\frac{du\,dv}{(\frac{r}{f})^2}\right] dx_T\,dy_T. \tag{4.16}$$

Notice in this case that the inner integral, enclosed in square brackets, is in fact independent of (x_T, y_T) and is equal to the area, A_0, of the lens aperture over $(\frac{r}{f})^2$, the square of the magnification. Therefore Eq. 4.16 can be further rewritten as

$$P_{D=\infty}(r) = \frac{\epsilon(r)\Upsilon_0 A_0}{(\frac{r}{f})^2}\int_{-\infty}^{\infty}\int_{-\infty}^{\infty} \Upsilon(x_T, y_T, r)\,dx_T\,dy_T. \tag{4.17}$$

By applying the normalisation of transmitted power expressed in Eqs. 4.6 and 4.17 can be reduced to

$$P_{D=\infty}(r) = \frac{\epsilon(r)A_0}{(\frac{r}{f})^2}. \tag{4.18}$$

Then, by substituting Eqs. 4.14 and 4.18 into Eq. 4.7, the overlap $O(r)$ can be expressed as

$$O(r) = \frac{\Upsilon_0}{A_0}\int_{-\infty}^{\infty}\int_{-\infty}^{\infty} \Upsilon(x_T, y_T, r) \times \left[\int_{-\infty}^{\infty}\int_{-\infty}^{\infty} A_R(u + x_T, v + y_T)D(\frac{u}{(\frac{r}{f})}, \frac{v}{(\frac{r}{f})})\,du\,dv\right] dx_T\,dy_T. \tag{4.19}$$

It is important to note that by the manipulations described here, dependence on the function $\epsilon(r)$, the function relating intensity in the scattering plane to intensity incident on the receiver objective, has been removed from the expression for overlap. The overlap is therefore purely a function of the optical geometry and the normalised

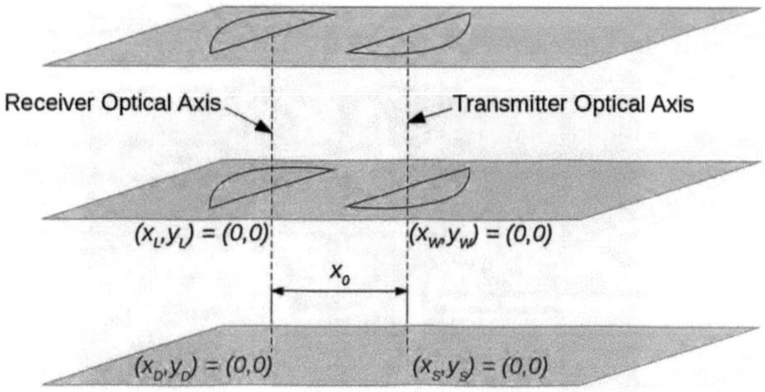

Fig. 4.4 Coordinates of transmitter focal plane relative to receiver focal plane

laser power distribution. The final step necessary for a full analytical derivation is an explicit expression of this distribution.

The function $\Upsilon(r)$ describes the laser intensity distribution on the scattering plane at range r. A similar approach to determining the receiver region in (x_T, y_T) can be used to calculate $\Upsilon(x_L, y_L, r)$, the intensity in the object plane due to the transmitter, by the convolution

$$\Upsilon(x_T, y_T, r) = \int_{-\infty}^{\infty}\int_{-\infty}^{\infty} A_T(\frac{r}{f}x_S + x_0 + x_T, \frac{r}{f}y_S + y_T)L(x_S, y_S)\, dx_S\, dy_S. \quad (4.20)$$

Here $L(x_S, y_S)$ is the near field output of the laser source observed in the focal plane of the transmitter objective and x_0 is the offset of the receiver optical axis from the transmitter optical axis. The coordinate shift from receiver optical axis to transmitter optical axis shown in Fig. 4.4 accounts for their offset in the overlap calculation and can be expressed by

$$(x_D, y_D) - (x_0, 0) = (x_S, y_S). \quad (4.21)$$

This is used to allow the laser emitter to be described around the transmitter optical axis (0, 0). The same shift is applied to assign coordinates (x_W, y_W) in the lens plane, where

$$(x_L, y_L) - (x_0, 0) = (x_W, y_W). \quad (4.22)$$

Since the far-field output of the laser is elliptical and approximately Gaussian it does not fill the aperture of the lens uniformly. In this work it is assumed that the far field laser output has an elliptical Gaussian distribution which is blocked or transmitted in the plane of the objective according to the lens aperture. Following the same approach used to define the receiver aperture as in Eq. 4.8 but here superimposing the elliptical Gaussian distribution of the laser, the effective aperture of the transmitter

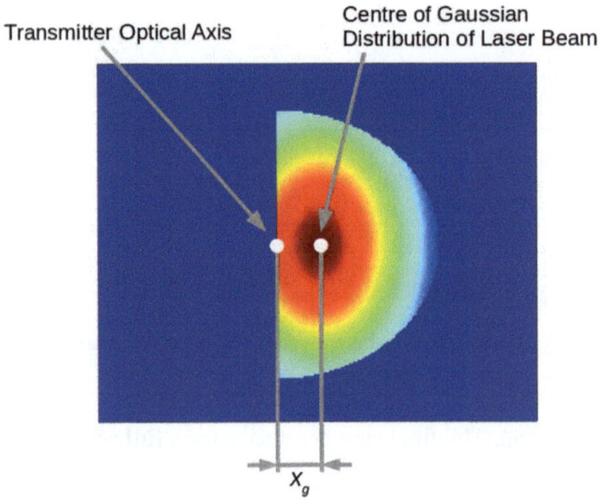

Fig. 4.5 Distribution of laser intensity across transmitter half-lens

$A_T(x_W, y_W)$ can be written as

$$A_T(x_W, y_W) = \begin{cases} e^{-\frac{1}{4}\left[\left(\frac{x_W - x_g}{\sigma_x}\right)^2 + \left(\frac{y_W}{\sigma_y}\right)^2\right]} & x_W^2 + y_W^2 < (\frac{d_L}{2})^2,\ x_W > 0 \\ 0 & \text{otherwise.} \end{cases} \tag{4.23}$$

Here $\sigma_x = 30\,\text{mm}$, $\sigma_y = 51\,\text{mm}$, and the center of the beam is offset x_g from the optical axis of the transmitter. Note that for this system x_g was set to 20 mm by tilting the laser beam to best fit the transmitter lens as shown in Fig. 4.5.

Finally, the laser diode used in this case is a single-stack device with three active regions. Accordingly, the near field laser output described is modelled by the indicator function $L(x_S, y_S)$ such that

$$L(x_S, y_S) = \begin{cases} 1 & |x_S| < \frac{l}{2},\ \frac{-w}{2} < y_S < \frac{+w}{2} \\ 1 & |x_S| < \frac{l}{2},\ \frac{-w}{2} < y_S - d < \frac{+w}{2} \\ 1 & |x_S| < \frac{l}{2},\ \frac{-w}{2} < y_S + d < \frac{+w}{2} \\ 0 & \text{otherwise,} \end{cases} \tag{4.24}$$

where the parameters $l = 0.235\,\text{mm}$, $w = 0.001\,\text{mm}$, and $d = 0.005\,\text{mm}$ are the length, width and separation distance of the active regions, respectively, taken from the manufacturer specification.

The function $\Upsilon(x_T, y_T, r)$ can now be calculated using Eq. 4.20 and substituted into Eq. 4.19 to determine the geometric overlap function. In order to calculate the overlap function $O(r)$, the integrals are computed numerically using discrete two-dimensional arrays. If the receiver aperture area A_0 is normalised to 1 and the laser power normalisation constant Υ_0 is used to set the total cross-sectional laser power to 1, the overlap $O(r)$ is unity when full overlap has been reached. Note that if the receiver area is not normalised the result is the expression for effective area rather than the overlap, but this is simply a scaled version of the overlap function.

4.3.2 Advantages of the Convolution Method

One advantage of this method is the fact that the need for analytical determination of the range-dependent cross-sectional overlap area $\mathcal{A}(r)$ (shown in Fig. 4.1) is removed, since the transmitted laser power cross section function and the receiver sensitivity function are multiplied and integrated with infinite limits of integration. The method does not include a straightforward way of accounting for displacement of the detector from the focal plane as the method of Halldórsson and Langerholc does. This was not deemed necessary for the lidar instrument characterised in this work as the detector is always placed in the focal plane. It has the additional advantages, however, of being extremely flexible and easy to adjust for different detector aperture shapes and sizes, different lens aperture shapes, sizes, and horizontal displacements, and different multi-mode laser diode near-field distributions.

4.3.3 Comparison of Overlap Calculation Methods

In order to validate the described method and its computation, effective area (overlap times receiver area) for an example biaxial lidar setup was calculated by the convolution method and compared with the results given for the same system in the literature. This system was first used for analytical calculation by Halldórsson and Langerholc [8] and later used for ray-tracing verification by Velotta et al. [13]. The system configuration considered, calculated for two different fields of view, is described in Table 4.1. Note that a uniform laser distribution was applied in all these calculations, and the laser beam was considered as a uniform projecting cone; the convolution approach therefore needed only be applied to the receiver.

Table 4.1 Comparison lidar parameters

Primary receiver entrance aperture radius	175 mm
Effective focal length	4,600 mm
Uniform laser beam radius	10 mm
Half-angle laser divergence	0.5 mrad
Offset between optical axes	200 mm
Receiver field of view	(A) 1 mrad and (B) 0.5 mrad

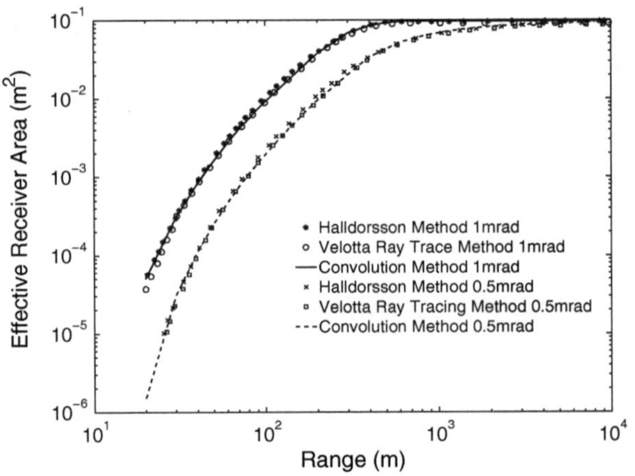

Fig. 4.6 Comparison of overlap calculation by the convolution method with results given by Halldórsson and Langerholc [8] and Velotta et al. [13] for an example biaxial system. The actual area of the receiver objective is 0.0962 m^2

The calculation methods are compared in Fig. 4.6. All three methods show close agreement for both fields of view. Note that in configuration B where the detector field of view is the same as the laser divergence, the effective area only approaches the actual receiver area (0.0962 m^2) at great distances.

4.3.4 Overlap Comparison for Different Optical Designs

In order to explore the overlap characteristics of different optical configurations, the calculated overlap functions of a number of possible prototype designs were compared using the geometrical method discussed. This is a similar exercise to one described in Chap. 3, but here the half-angle field of view was held constant at 0.75 mrad, and the effect of an additional parameter, transmitted laser beam tilt, was considered for the prototype design. The detector, the laser, and the focal length of the lenses were kept the same for all configurations. The overlap of each design is shown in Fig. 4.7.

The thin solid line shows the coaxial common optics system where one full 150 mm diameter lens is shared by the transmitter and receiver channels. This configuration has the quickest turn-on, but it does not offer the optical isolation benefits of a biaxial system. The other extreme is a biaxial system using two full 150 mm lenses with 21 mm between their edges and therefore 171 mm offset between their optical axes. This is shown by the bold dash-dot line. The second slowest overlap function, shown by the thin dash-dot line, was calculated for two full 106 mm diameter lenses with optical axes offset by 127 mm; each of these lenses has the same area as a 150 mm half-lens. Shown by the bold dashed line is the divided lens system with no

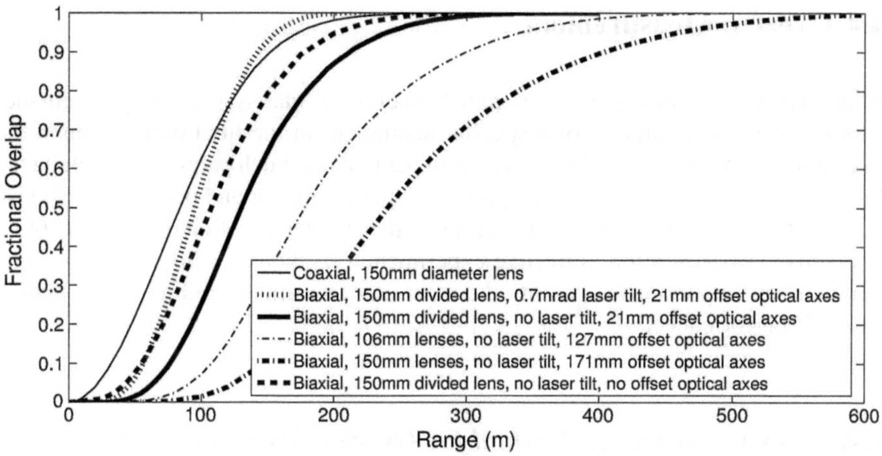

Fig. 4.7 Calculated overlap function for various lidar prototype designs. Half-angle FOV is 0.75 mrad for all designs considered

offset between optical axes. This configuration would allow no room to fit the laser and detector at the focal points without some additional complexity in the design, but it is shown as a reference to highlight the differences between this configuration, the coaxial common optics system, and the 21 mm offset half-lens prototype design shown by the bold solid line. Considering the way the laser beam and receiver field of view diverge from a lidar instrument, it is not surprising that the zero offset split lens has slightly slower overlap than the shared-lens coaxial arrangement. It is interesting to see, however, that the overlap function of the zero offset lens sits approximately half way between these other two.

The other comparison to note is that between the prototype, shown by the bold solid line, and the same system with a transmitted laser beam tilt of 0.7 mrad toward the receiver, indicated in Fig. 4.7 by the thin hatched line. Laser beam tilt can be achieved by shifting the laser slightly away from the optical axis of the transmitter. Though the perpendicular alignment of the laser beam is more straightforward in a production setting, the laser could be tilted to achieve an overlap function closer to that of the coaxial common optics configuration if additional close range sensitivity were desirable. Note, however, that before changing the optical design a thorough analysis of the multiple scattering effects of fog and low clouds would be recommended, since for ceilometer applications the ability to detect the presence of these features via multiple scattering is more important than a precise inversion of extremely close range aerosol extinction. A small but detectable signal that allows fog and low cloud features to be detected without saturation is desirable at very close ranges to the instrument so that the gain does not have to be dropped to compromise detection of higher features in the atmosphere. In addition, aberration effects of shifting the laser away from the optical axis would need to be considered before implementing this change.

4.4 Overlap Measurement

While theoretical techniques are extremely useful for lidar system design, their success for overlap calibration of a specific instrument in the field depends upon the availability of reliable knowledge of system parameters. Such parameters include the laser source power distribution and propagation, the angular and positional response of the detector and the performance of the optics, which are often difficult to obtain in practise [18]. Therefore, effective experimental methods for lidar overlap determination are important for gaining the most accurate understanding of the optical system's performance.

4.4.1 A Novel Imaging Method for Overlap Measurement

After the overlap of the prototype system was calculated, it was measured using a novel imaging technique. The method proposed herein for the measurement of lidar overlap is similar in concept to placing a planar target with isotropic scattering properties at various distances from the instrument and measuring the lidar return from each distance. This method is often impractical, however, since a long (up to several kilometer) unobstructed path with homogeneous atmospheric properties is required, and a large target is needed at the far end of the measurement range. To overcome these practical difficulties, an optical system is used to present a virtual image of the target to the instrument at various ranges. The simplest configuration that is required to achieve this is presented in Fig. 4.8.

In this configuration an imaging lens outside the lidar system is used to present a virtual image of a scattering target to the instrument. Using the Gaussian lens law [22]

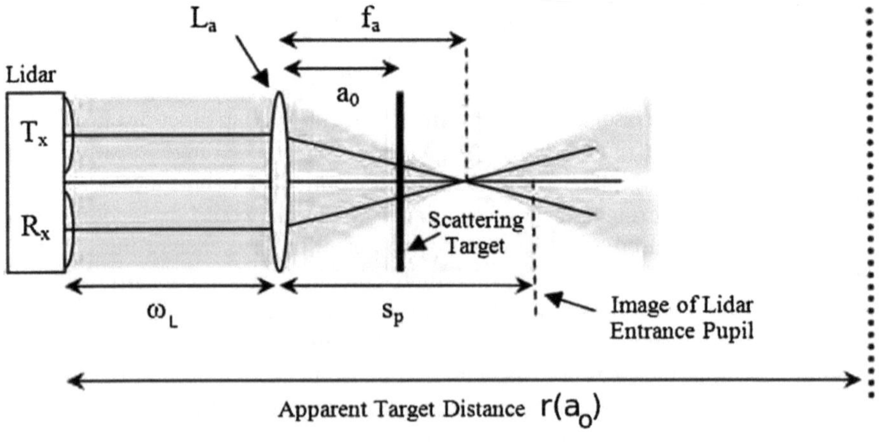

Fig. 4.8 Imaging system for determination of overlap

it is straightforward to show that this image appears at a distance, $r(a_o)$, from the instrument given by

$$r(a_o) = \omega_L + \left| \frac{f_a a_o}{a_o - f_a} \right|, \tag{4.25}$$

where the constant ω_L is the distance from the lidar to the imaging lens, f_a is the focal length of the imaging lens, and a_o is the distance from the imaging lens to the target. It is clear that when the target is in the focal plane its image is presented at infinity. It is important to realise that although $r(a_o)$ is the distance at which the image of the target is presented to the instrument it cannot be simply substituted as range into the geometric factor $G(r)$ in Eq. 4.2. This is because, as noted previously, the light collected by the lidar is proportional to the inverse square of the distance from the target to the entrance pupil of the instrument. The insertion of an imaging lens between the target and the instrument has effectively moved the position of the entrance pupil.

Using the Gaussian lens law again it is straightforward to show that the (real) image of the entrance pupil is located at a distance, s_p, from the imaging lens, given by

$$s_p = \frac{f_a \omega_L}{\omega_L - f_a}. \tag{4.26}$$

Consequently, the distance, $r_p(a_o)$, from the target to the entrance pupil is given by

$$r_p = \left| s_p - a_o \right|. \tag{4.27}$$

In order to measure the overlap function it is necessary to make two assumptions. First, it is assumed that the target is uniform with isotropic scattering efficiency. Secondly, it is assumed that over the short target translation distances in a controlled laboratory environment, variations in atmospheric transmission losses due to target translation are negligible. With these assumptions the terms $K_s \beta(r) T(r)$ in Eq. 4.1 can be replaced by a constant, $\frac{1}{C}$, and using Eq. 4.3 the overlap can be written,

$$O\big(r(a_o)\big) = CP\big(r(a_o)\big)\big(r_p(a_o)\big)^2, \tag{4.28}$$

where the apparent target distance, $r(a_o)$, and the apparent distance from the target to the entrance pupil, $r_p(a_o)$, are given explicitly by Eqs. 4.25 and 4.27 respectively.

Although the basic imaging system of Fig. 4.8 provides a way to measure the overlap function of a lidar instrument in a confined space, it is not easy to use in practise. Because the distance $r(a_o)$ is a non-linear function of the target distance a_o, measurements are clustered around the focal plane of the imaging lens. Furthermore, for the characterisation of typical lidar systems by this method, the range correction factor $\big(r_p(a_o)\big)^2$ depends much more strongly on the target distance than the overlap function does, and thus small errors in the measurement of target distance a_o can

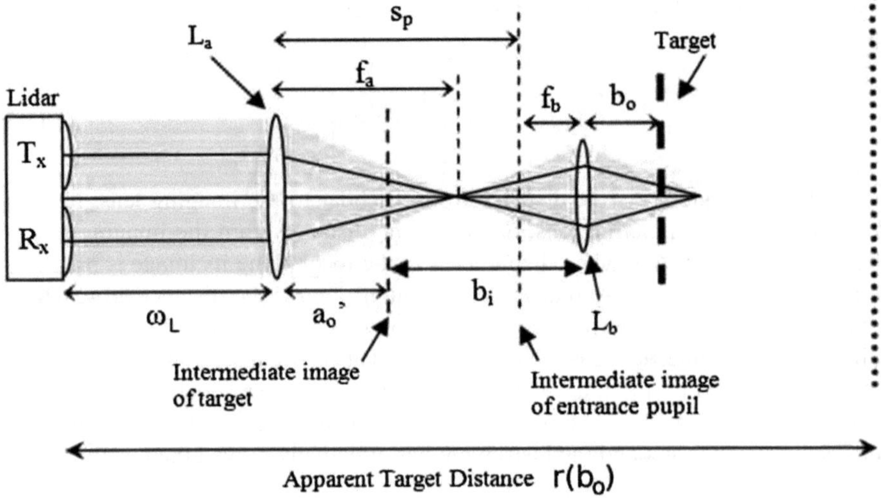

Fig. 4.9 Compound imaging system for determination of overlap

introduce substantial errors in the overlap measurement. The second of these problems and to some extent the first may be overcome by using a second imaging lens to form the compound imaging system shown in Fig. 4.9.

In essence, lens L_b acts as a relay that presents a real image of the target to lens L_a (this image becomes the object for L_a). The main advantage of this system over the first is the fact that here the distance between the lenses can be adjusted in order to form the image of the entrance pupil of the lidar at infinity, thereby removing the need for range correction.

Again employing the Gaussian lens law the virtual range, here $r(b_o)$, can be calculated using Eq. 4.25 by

$$r(b_o) = \omega_L + \frac{f_a a_o'}{f_a - a_o'}, \tag{4.29}$$

where

$$a_o' = f_b + s_p - \frac{b_o f_b}{b_o - f_b}. \tag{4.30}$$

The geometric factor from Eq. 4.28 can now be formulated for this system with the added relay lens. In this case the image of the entrance pupil is formed at infinity, so as b_o changes, the solid angle subtended by the entrance pupil does not change. The range dependence of the geometrical factor is thereby removed, allowing Eq. 4.28 to be formulated as

$$O(r(b_o)) = CP(r(b_o)), \tag{4.31}$$

where the distance to the virtual image of the target, $r(b_o)$, is given explicitly by Eq. 4.30.

Note that this expression with constant C only applies if the system being characterised reaches full overlap at some range. Otherwise C can not be used to set the maximum value of $O(r)$ equal to 1. However, since this method images to infinity, it should be clear from the result whether or not the slope of the overlap function reaches zero at some point after overlap onset. The slope should reach zero at the full overlap range and then stay at zero out to infinite imaging distance. If the overlap reaches a maximum and then decreases, misalignment of the system is revealed. In this case, the overlap at the maximum cannot be considered to be unity and measurements from the system cannot be inverted properly, but the misalignment has been identified, and the signal intensity could still be overlap-corrected in a relative sense. The ability of this method to present virtual clouds at infinite ranges therefore gives it a significant advantage over other methods for elastic lidar overlap measurement, since it can reveal small misalignments in the lidar system under test that might result in full overlap not ever being reached by that system.

A compound imaging system similar to that described above was employed to measure the overlap of the prototype lidar with a single-stack laser. In the following section the experimental setup is described and the results of the measurement given in comparison with the calculated results and results from horizontal hard target measurements on an airstrip.

4.4.2 Experimental Determination of Prototype Lidar Overlap

The experimental setup used to measure the lidar overlap function by application of the technique described in the previous section has the same basic layout as the configuration presented in Fig. 4.9. It is clear, however, that in order for the system to provide an accurate measure of overlap, the aperture of any optics used must not restrict the system aperture and any aberrations introduced by additional components should have a negligible effect. In this work, the large imaging lens L_a of the compound imaging system of Fig. 4.9 was replaced by a 0.25 m diameter spherical mirror with a focal length $f_a = 3.057$ m. Consequently the mirror was tilted at an angle of 1.5° to fold the imaging system as shown in Fig. 4.10, placing a 3° angle between the light directly from the lidar and the light reflected by the mirror.

The relay lens was a 75 mm diameter achromatic doublet of nominal focal length $f_b = 0.4$ m (Edmund Optics NT45-419). As the calculation of virtual target distance is quite sensitive to this focal length, careful determination of the focal length of the relay lens was necessary and this was found to be $f_b = 0.402$ m at a wavelength of 905 nm. The 1.5° angle was chosen to be as small as possible to reduce aberration but sufficient to allow the lidar beam to pass unobstructed to the mirror. The gap between the lidar and the mirror, ω_L, was set to 20.05 m to ensure that the returned signal could be separated from electronic noise generated by the laser pulse. The distance between the mirror and the relay lens was adjusted such that the image of

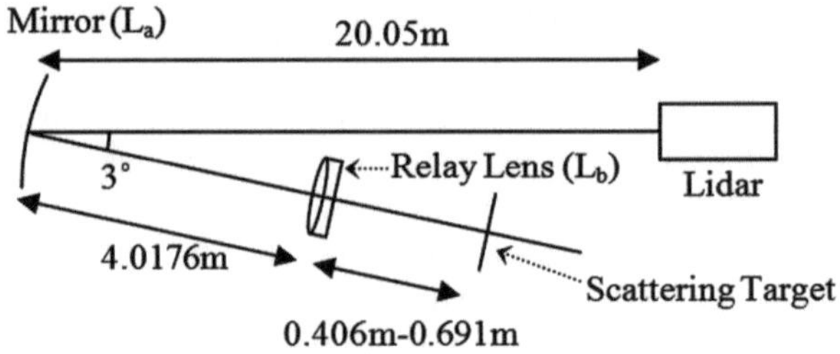

Fig. 4.10 Overlap measurement experimental set-up

the lidar entrance pupil as viewed from the target was located at infinity, and was therefore equal to 4.0176 m, the sum of the entrance pupil image distance, s_p, and the relay lens focal length, f_b.

To ensure that aberrations were negligible the system was modeled using OSLO optical design software. It was found that at the most aberration-sensitive point, where the target is placed such that its virtual image is set at infinity, the calculated minimum radius of the focused spot image of an ideal collimated laser beam was 0.031 mm (approximately twice that of the diffraction-limited value). At this target distance (well past the point of full overlap such that laser stripe image should be well within the detector image) the radius of the image of the laser stripe was calculated to be 0.77 mm and the radius of the image of the detector was calculated to be 1.64 mm. The influence of the 0.031 mm minimum spot radius should therefore be insignificant, increasing the laser image radius by 4 % and the detector image radius by 2 %. Thus the level of aberration present in the imaging system was judged to be acceptable.

A matt-black painted aluminium target and a white paper target were used for this work. It was assumed that these targets provided scattered returns that were largely isotropic over the range of accepted target angles. At each measurement point the difference between the returns from the white and black targets was measured to compensate for any spurious signals arising from scattering within the imaging system. In all measurements the laser was operated at normal operating power to ensure its beam profile matched that used in the field, and lidar returns were measured from each target at each distance. Because the signals returned were significantly larger than those obtained from clouds, no reverse voltage was applied to the APD, so it behaved as a simple photodiode. This mode of operation was not expected to change the overlap characteristics of the lidar system. The target was mounted on a precision motorized translation stage and measurements were taken at target positions ranging from 0.406 to 0.691 m from the back surface of the relay lens which corresponded to lidar ranges from 17 to 8,402 m.

The results of overlap measurements made using this virtual imaging technique are shown in Fig. 4.11 as red asterisks. These are compared with the overlap calculated by the geometric calculation described earlier in this chapter, shown as a bold line. An additional measurement of the overlap was made by gathering hard target

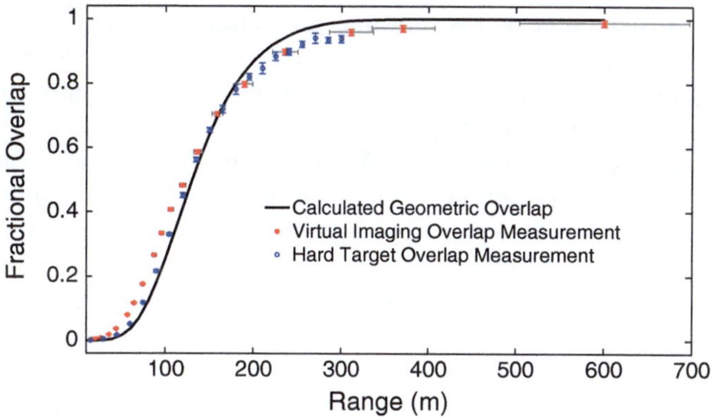

Fig. 4.11 Calculated and measured overlap

measurements at ranges up to 300 m with the lidar, at a low gain setting, pointing horizontally along a retired local air field on a windy afternoon when aerosol content was assumed to be low. This comparison measurement was made up to a range of 300 m using a target of roughly 1.5 m × 1.5 m. These results, range-corrected, are shown in Fig. 4.11 as blue circles.

The data set gathered by the virtual imaging technique was normalised to 1 at 8,402 m, essentially infinity, and the range-corrected horizontal hard target measurements were in turn normalised to the virtual imaging data at 240 m. Since alignment for the hard target measurements became difficult at farther ranges, 240 m was chosen to ensure that the laser beam was not approaching the edge of the target. If the laser beam had inadvertently moved slightly off target at the normalisation point, the results would have been skewed.

The largest potential sources of error in the imaging method are the focal length of the relay lens (which was measured in the laboratory at 633 nm and then recalculated at 905 nm based on the lens design using OSLO), the focal length of the mirror (also measured in the laboratory) and the determination of the distance from the target to the relay lens at each measurement point. Error in either of the focal length measurements will shift the virtual target distance to stretch or compress the S shape of the typical overlap function, as will systematic error in the measurement of target position. Uncertainty in each of these three parameters was estimated to be 0.1 %. At a range of 150 m (approximately half the distance to full overlap for this system, at a calculated overlap of 0.64), a 0.1 % error in the focal length measurement of the lens would shift the virtual target distance 4 m, while a 0.1 % error in either the focal length measurement of the mirror or the measurement of target distance would shift the virtual target distance by 3 m at the same point. Less shift at closer ranges and more shift at farther ranges is expected due to the fact that the virtual target distance, $r(b_o)$, is a nonlinear function of distance b_o from the relay lens to the target. The uncertainty in calculated range $r(b_o)$ due to ± 0.1 % uncertainties in the focal lengths and target distances was computed as a root mean square sum of the effects of the individual uncertainties at each range and is shown as grey error bars along the x-axis. Error bars along the y-axis indicate, for each measurement technique, uncertainties of ±2 standard deviations due to random fluctuations in the power measurements that were used to derive the fractional overlap.

Beyond ranges of approximately 125 m (overlap of 50 % and greater) the the hard target measurements and the virtual imaging measurements agree within the calculated uncertainties. However, regardless of possible adjustment of the point selected for normalisation of the hard target data, there is a significant disagreement below 125 m due to error that has not been accounted for in the data analysis. There are a number of possible sources of this error. Alignment of the prototype could have shifted slightly due to vibration during its transport from the laboratory to the air field or as a result of possible small operating temperature differences between the two measurement environments. Error could also have resulted from one or both of two assumptions made in this work: the assumption that the target scattering is isotropic over the accepted angle of the optical system, and the assumption that the spatial response of the APD detector is the same regardless of bias voltage.

Since the optical system used in the calculation was an ideal system, the measured overlap was not expected to agree exactly with the calculation. The assumption, for the purposes of calculation, of aberration-free optics in the lidar prototype, is of course a tenuous one. The earlier onset of the experimental data sets could possibly be explained by aberrations, incidental scattering within the optics, or other optical effects in the lidar system. The fact that both of the experimental data sets take longer to reach full overlap than the calculated curve suggests a possible small, unintentional tilt of the laser away from the receiver. As mentioned previously, diffraction effects of the laser source can have considerable influence on the overlap function of a lidar system with an annular beam shape. Since the beam shape of this system is not annular and since the objective aperture is much larger than the laser wavelength, the assumption of a Gaussian laser distribution in the calculation was assumed to be reasonable and diffraction effects were not expected to contribute significant error beyond the more substantial errors likely resulting from other factors such as uncharacterised aberrations.

4.5 Conclusions

An analytical expression based on geometric optics has been derived specifically for the prototype lidar optical system. Overlap calculated using this method was validated by comparison with an analytical calculation and a ray-tracing method for a system described in the literature. While it was designed specifically for the task of characterising the overlap function resulting from the unique geometry of the prototype optics, this method has a benefit of straightforward adaptability for a wide variety of aperture arrangements, emitter shapes, and laser distributions, and it proved to be a useful tool for optimising the optical design of the prototype.

Following the calculation of overlap, a compound imaging system for the measurement of lidar overlap in the laboratory was designed and proved. In essence, the method presents to the lidar instrument a virtual image of a scattering target at a specified range, and the response is measured. By using a compound lens system it is possible ensure that the entrance pupil of the instrument as seen from the scattering target is presented at infinity. In this case, the signal returned to the lidar is not diminished by the inverse square characteristic and becomes a direct measure of the overlap function.

Hard target overlap measurements used to validate overlap measured using this imaging system fell within the calculated uncertainty beyond the range of 50 % overlap. Discrepancies at closer ranges likely resulted from one or more of a number of possible sources of error that were identified but could not be quantified for this experiment. At ranges corresponding to about 50–80 % overlap (125–200 m), the calculated overlap fell within the error bars on the overlap determined using the imaging method. However, below and above this region, the calculations did not fall within the measurement error. Discrepancies between theoretical and experimental determinations of overlap are to be expected and highlight the importance of measuring the overlap. Here these discrepancies probably arise from poorly understood optical properties of the lidar system that are not accounted for in the calculation.

Overlap is influenced by a number of factors that are either unknown or difficult to measure in practise and therefore any theoretical determination of overlap is likely to be flawed in some way. For example, the far field distribution of the laser, which may be poorly understood, influences the overlap considerably. Similarly, the angular response of the detector, which was assumed to be uniform in this study, is likely to vary in practise, especially when combined with a bandwidth-restricting interference filter. This again will modify the effective aperture of the receiver and in turn affect the overlap function. Measurement of the overlap function is therefore critical if data from the instrument near and below the full overlap distance is to be used.

Note that regardless of the technique used to calculate or measure overlap, significant multiple scattering at close ranges can produce significantly greater returns than would be expected from single-scattering alone [12]. Therefore the effects of multiple scattering, in addition to the overlap profile, should be taken into consideration when lidar signals are evaluated.

If reliable estimates of aerosol distributions or visibility are required at close ranges then it is usually necessary to calibrate the lidar in at least part of overlap region. The measurement system discussed here shows that this is indeed possible in a controlled laboratory environment.

References

1. Y. Sasano, H. Shimizu, N. Takeuchi, M. Okuda, Geometrical form factor in the laser radar equation: an experimental determination. Appl. Opt. **18**(23), 3908–3910 (1979)

2. U. Wandinger, in *LIDAR: Range-Resolved Optical Remote Sensing of the Atmosphere*, ed. by C. Weitkamp. Introduction to lidar (Springer, New York, 2005), pp. 1–18
3. K. Stelmaszczyk, M. Dell'Aglio, S. Chudzynski, T. Stacewicz, L. Wöste, Analytical function for lidar geometrical compression form-factor calculations. Appl. Opt. **44**(7), 1323–1331 (2005)
4. R. Measures, *Laser Remote Sensing: Fundamentals and Applications* (Wiley-Interscience Publication, Wiley, New York, 1984)
5. E.M. Patterson, D.W. Roberts, G.G. Gimmestad, Initial measurements using a 1.54-μm eyesafe raman shifted lidar. Appl. Opt. **28**(23), 4978–4981 (1989)
6. R.E.W. Pettifer, G.J. Jenkins, P.G. Healey, J.H. Convery, A large coaxial lidar for elastic and inelastic scattering studies of the stratosphere. Opt. Quant. Electron **8**, 409–423 (1976). doi:10.1007/BF00624832
7. C. Münkel, N. Eresmaa, J. Räsänen, A. Karppinen, Retrieval of mixing height and dust concentration with lidar ceilometer. Bound-Lay Meteorol **124**, 117–128 (2007). doi:10.1007/s10546-006-9103-3
8. T. Halldórsson, J. Langerholc, Geometrical form factors for the lidar function. Appl. Opt. **17**(2), 240–244 (1978)
9. J. Harms, W. Lahmann, C. Weitkamp, Geometrical compression of lidar return signals. Appl. Opt. **17**(7), 1131–1135 (1978)
10. G.M. Ancellet, M.J. Kavaya, R.T. Menzies, A.M. Brothers, Lidar telescope overlap function and effects of misalignment for unstable resonator transmitter and coherent receiver. Appl. Opt. **25**(17), 2886–2890 (1986)
11. H. Kuze, H. Kinjo, Y. Sakurada, N. Takeuchi, Field-of-view dependence of lidar signals by use of Newtonian and Cassegrainian telescopes. Appl. Opt. **37**(15), 3128–3132 (1998)
12. K. Sassen, G.C. Dodd, Lidar crossover function and misalignment effects. Appl. Opt. **21**(17), 3162–3165 (1982)
13. R. Velotta, B. Bartoli, R. Capobianco, L. Fiorani, N. Spinelli, Analysis of the receiver response in lidar measurements. Appl. Opt. **37**(30), 6999–7007 (1998)
14. P. Kokkalis, G. Georgoussis, A. Papayannis, D. Hatzidimitriou, J. Porteneuve, R. Mamouri, G. Tsaknakis, in *Proceedings of the 8th International Symposium on Tropospheric Profiling*, ed. by A. Apituley, H.W.J. Russchenberg, W.A.A. Monna. Optimization-through optical design-of a multi-wavelength fiber-based Raman-lidar system in the near-field for vertical aerosol measurements in the troposphere, vol. paper S06–P01 (ISTP, 2009)
15. I. Berezhnyy, A combined diffraction and geometrical optics approach for lidar overlap function computation. Opt. Laser Eng. **47**(7–8), 855–859 (2009)
16. K. Tomine, C. Hirayama, K. Michimoto, N. Takeuchi, Experimental determination of the crossover function in the laser radar equation for days with a light mist. Appl. Opt. **28**(12), 2194–2195 (1989)
17. S.W. Dho, Y.J. Park, H.J. Kong, Experimental determination of a geometric form factor in a lidar equation for an inhomogeneous atmosphere. Appl. Opt. **36**(24), 6009–6010 (1997)
18. U. Wandinger, A. Ansmann, Experimental determination of the lidar overlap profile with Raman lidar. Appl. Opt. **41**(3), 511–514 (2002)
19. S. Hu, X. Wang, Y. Wu, C. Li, H. Hu, Geometrical form factor determination with Raman backscattering signals. Opt. Lett. **30**(14), 1879–1881 (2005)
20. C. Raman, K. Krishnan, A new type of secondary radiation. Nature **121**, 501–502 (1928)
21. J.L. Guerrero-Rascado, M.J. ao Costa, D. Bortoli, A.M. Silva, H. Lyamani, L. Alados-Arboledas, Infrared lidar overlap function: an experimental determination. Opt. Express, **18**(19), 20350–20359 (2010)
22. W.J. Smith, *Modern Optical Engineering* (SPIE, New York, 2008)

Chapter 5
Determination of Cloud Base Height and Vertical Visibility from a Lidar Signal

5.1 Introduction

The instrument described in Chap. 3 was designed to automatically report cloud height and vertical visibility. This chapter presents the signal processing methods that have been applied to the prototype and subsequently evaluates their performance. While ceilometers are widely available for purchase, the methods by which cloud base height is determined from their signals is usually kept as proprietary commercial information, even though this information can be important to users in the scientific community. Here, a comprehensive cloud detection signal processing method for a ceilometer is described in detail.

This chapter begins with a discussion of signal conditioning, and then describes calibration of attenuated backscatter output and provides some example measurements of this parameter. Following this, cloud base height definitions in the literature are reviewed, and then the novel two-part automated algorithm for the determination of cloud base height from measured signals that has been developed for the prototype is described. The performance of the instrument is then compared with that of a research ceilometer at the Chilbolton Observatory. Finally, since ceilometers are expected to report visibility when no cloud base is detected, the implementation of a standard vertical visibility algorithm is described and evaluated. Note that aside from the prototype data, all other data from the Chilbolton Observatory used in this chapter, specifically the CT75k ceilometer data, the cloud camera images, and the rainfall rate and ground-based visibility data, were provided by the NERC-funded Chilbolton Facility for Atmospheric and Radio Research (CFARR).

5.2 Impulse Response Correction

Before feature detection methods are applied to a signal, any significant artefacts of the detection process should be removed. In a pulsed system such as the prototype, fast recovery from an impulse is an important feature of its amplifier [1]. Even in

J.D. Vande Hey, *A Novel Lidar Ceilometer*, Springer Theses,
DOI 10.1007/978-3-319-12613-5_5

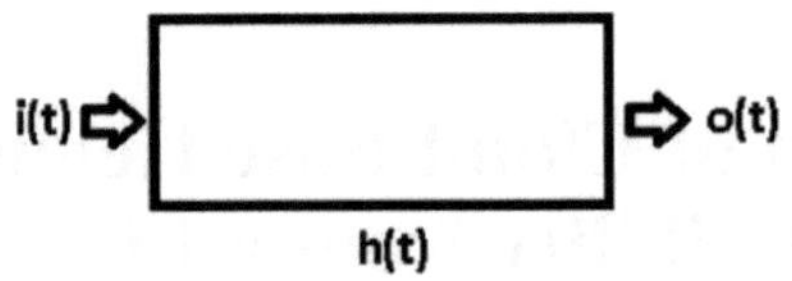

Fig. 5.1 Amplifier impulse response diagram

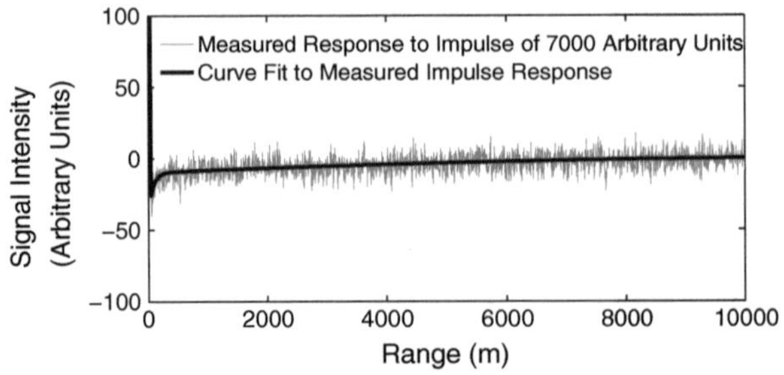

Fig. 5.2 Amplifier impulse response characterisation

a system with a fast recovery, however, artefacts may still appear in the signal. The impulse response behaviour of an amplifier is represented schematically in Fig. 5.1, where $i(t)$ is the input signal, $h(t)$ is the impulse response of the amplifier, and $o(t)$ is the output signal.

In the prototype lidar, the AC-coupled amplifier used on the receiver exhibits an impulse response characteristic as shown in Fig. 5.2. The measured impulse to a 7×10^3 magnitude signal is shown as a thin line and the fitted function (exponential from bins 8–30) is shown as a bold line. This negative "kickback" of the signal following detection of electronic noise from the pulse and any close-range optical returns is due to a non-uniform response to the frequencies in the impulse. Once the impulse response has been measured and fitted to a smooth function, this function can be used to correct the response.

Given a filter with characterised time-domain response $h(t)$, the input signal $i(t)$ can be recovered from the output signal $o(t)$ as follows. First, $o(t)$ is expressed as the convolution,

$$o(t) = i(t) * h(t), \tag{5.1}$$

which, by application of the convolution theorem [2], becomes in the frequency domain

$$O(\omega) = I(\omega)H(\omega). \tag{5.2}$$

Therefore,

$$I(\omega) = O(\omega)/H(\omega), \tag{5.3}$$

where each of these frequency domain functions is the Fourier transform of the corresponding time domain function, defined such that

$$H(\omega) = \mathcal{F}[h(t)] = \int_{-\infty}^{\infty} h(t)e^{-j\omega t}dt. \tag{5.4}$$

The function $H(\omega)$ can be expressed as the product of a real magnitude $a(\omega)$ and a complex phase component $e^{i\phi(\omega)}$ with real phase angle ϕ as

$$H(\omega) = a(\omega)e^{i\phi(\omega)}. \tag{5.5}$$

By substituting this into Eq. (5.3), $I(\omega)$ can now be expressed as

$$I(\omega) = O(\omega) \cdot \frac{e^{-i\phi(\omega)}}{a(\omega)}. \tag{5.6}$$

The amplification electronics in the prototype act as a low-pass filter, with a fairly flat response at low frequencies and roll-off at higher frequencies following an approximately Gaussian distribution as shown in the upper graph in Fig. 5.3. The lower graph in Fig. 5.3 shows the phase response.

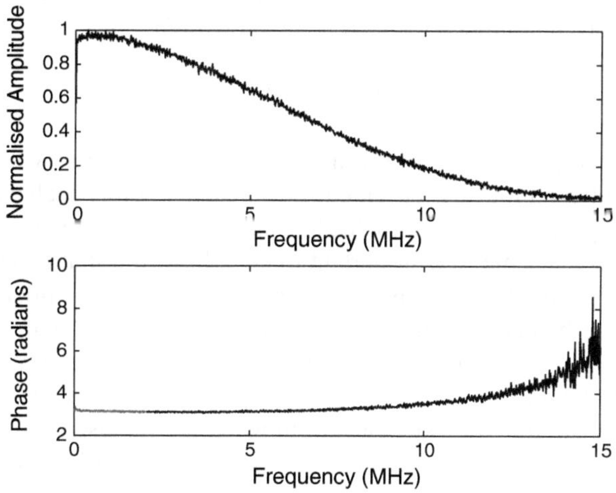

Fig. 5.3 Frequency response of prototype electronics subjected to an impulse. The *upper graph* shows the absolute amplitude response and the *lower graph* shows the phase response

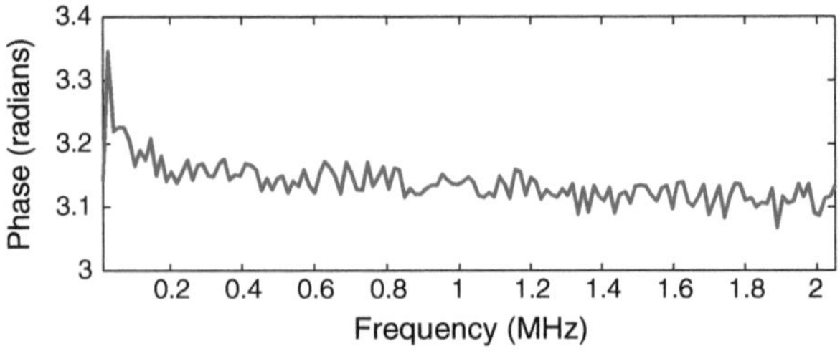

Fig. 5.4 Close-up view of phase of low frequency impulse response shown in the *lower graph* in Fig. 5.3

A full inverse filter would compensate for inconsistencies in both the amplitude and phase response of the electronics at all present frequencies. However, boosting high frequencies to achieve a flat amplitude response increases the noise in the signal substantially. Since the amplitude response is fairly flat at low frequencies and since the major phase distortion occurs at high frequencies which are naturally dampened by the system, it is actually the small phase distortion at the lowest frequencies shown close up view in Fig. 5.4 that was found to be the primary source of the negative "kickback" of the electronics following an impulse.

If it can be assumed that phase-only correction is sufficient, $a(\omega)$ in Eq. (5.6) can be set equal to 1 and the phase-corrected signal in the frequency domain can be expressed

$$I(\omega) = O(\omega) \cdot e^{-i\phi(\omega)}. \quad (5.7)$$

Then, by returning to the time domain through application of an inverse Fourier transform, the original input signal, $i(t)$, can be written

$$i(t) = \mathcal{F}^{-1}[O(\omega) \cdot e^{-i\phi(\omega)}]. \quad (5.8)$$

The impulse response, $h(t)$, of the amplifier can be measured while the system is running if an optical impulse is provided to the detector before each measurement. In this way, any drift in the electronics due to changes in gain, temperature, etc., can be accounted for in real time. It is important to note, however, that if saturation occurs during either the impulse response measurement or the actual lidar measurement, the method described here cannot be expected to correct the signal.

An example of a signal before and after impulse response correction using this technique is shown in Fig. 5.5, and a close up of the data along the x-axis is shown in Fig. 5.6. Even though the "kickback" following an impulse may be small, it can have major implications for inversion of the signal if it is not corrected, so this step is critical to the quality of results given by the instrument.

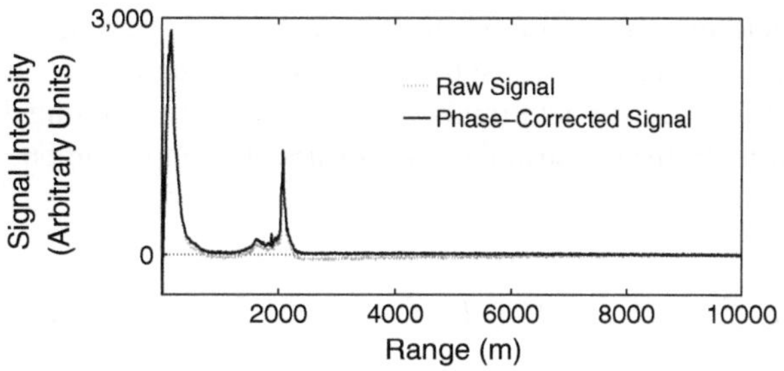

Fig. 5.5 Amplifier impulse response correction applied to data from 5 December, 2011

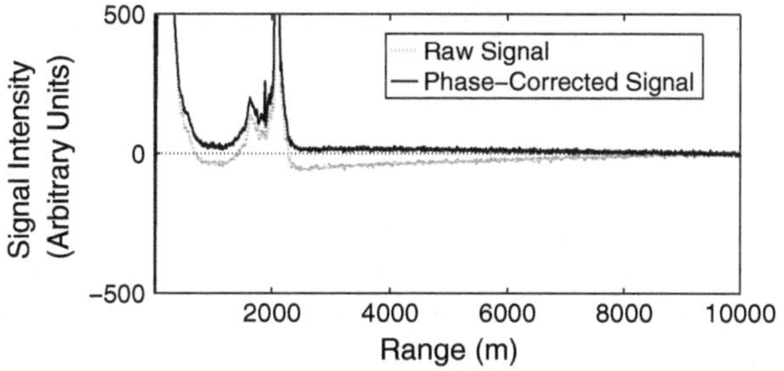

Fig. 5.6 Close up view of data shown in Fig. 5.5

5.3 Attenuated Backscatter Profile

Once the signal has been corrected for impulse response, it is ready to be processed. The first output that needs to be generated is the attenuated backscatter profile, that is, the calibrated, range-corrected signal. This is an important output for a ceilometer as it can be used to assess the signal and compare returns with those from other lidar instruments.

Recall from Chap. 2, that the attenuated backscatter profile $\beta'(r)$ at a given wavelength can be calculated as follows,

$$\beta'(r) = \frac{P(r)r^2}{P_0 \frac{\tau c}{2} O(r) A_0}, \tag{5.9}$$

where $P(r)$ is the power measured from range r, r^2 is the range-correction factor, P_0 is the transmitted laser power, τ is the pulse length (100×10^{-9} s for the prototype), c is the speed of light, $O(r)$ is the fractional overlap at range r, and A_0 is the area of the receiver objective aperture (0.0088 m^2 for the prototype). $O(r)$ is included here so that the attenuated backscatter profile can be estimated at ranges before full overlap is reached. Note that the wavelength dependence of $\beta(r)'$ has not been expressed in these equations under the assumption of a fixed, narrow laser wavelength typically used in lidar systems.

In order to determine P_0 and $P(r)$, the key factors for an absolute calibration, the efficiency of the optical system must be characterised and the total output power and absolute gain of the receiver determined. In a manufacturing setting these need to be checked for variation from instrument to instrument due to component tolerances.

5.3.1 Calculation of Transmitted and Received Power

5.3.1.1 Transmitted Laser Power

The laser output power leaving the instrument during a pulse P_0 can be determined such that

$$P_0 = P_L \eta_F \eta_O \eta_G \eta_A, \tag{5.10}$$

where P_L is the laser output power during the pulse, η_F is the transmission efficiency of the optical bandpass filter, η_O is the transmission efficiency of the objective, η_G is the transmission efficiency of the external glass plate of the instrument, and η_A is the fraction of the divergent laser beam that is transmitted by the objective aperture.

The laser output power P_L is measured by collecting the entire output of the laser. Before it is installed in the prototype lidar, the laser drive current is set to achieve an output, P_L, of 83 W at 40 C to meet the laser safety standards as discussed in Chap. 3. In the instrument a photodiode is included so that the laser power can be monitored and adjusted if necessary in order to maintain a stable value of P_0.

The transmission efficiency, η_F, of the optical bandpass filter that protects the laser from inadvertent exposure to direct solar radiation is specified by the manufacturer as being greater than 0.8 at 915 nm, the transmission efficiency, η_O, of the AR-coated objective lens is 0.98, and the transmission efficiency, η_G, of the external glass plate is 0.95.

The final factor, η_A, the fraction of the laser light that is collected by the aperture, is calculated by assuming a Gaussian distribution along each of the primary orthogonal axes of the propagating laser beam. Manufacturer-specified divergence angles are used to calculate the distribution at a distance away from the laser equal to the focal length of the lens. The semicircular aperture is superimposed on the laser distribution, and the fraction of energy allowed through is calculated by numerical integration over the aperture area.

5.3.1.2 Calibrated Received Power

In order to determine the range-dependent power $P(r)$ incident on the receiver, the receiver gain must be determined and the efficiency of the receiver characterised. $P(r)$ can be expressed as

$$P(r) = \frac{P_M(r)}{G_{A_2} G_{A_1} G_{APD} \mathcal{R} \eta_F \eta_O \eta_G}, \quad (5.11)$$

where $P_M(r)$ is the amplified output (volts) measured from range r. The three stages of amplification considered in the electronics are G_{A_2}, the gain (unitless) of the secondary amplifier, G_{A_1}, the gain (V/A) of the primary amplifier, and G_{APD}, the gain (unitless) of the APD itself. The constant $\mathcal{R}$ is the responsivity of the APD (0.52 A/W). The efficiencies η_F, η_O, and η_G are the same as those described for the laser channel. The gains G_{APD}, G_{A_1}, and G_{A_2} are variable, and since the APD gain varies significantly in a nonlinear fashion with both temperature and bias voltage, G_{APD} is calibrated for each measurement through the use of an LED with fixed power output. This calibration is particularly important because the APD gain needs to be adjusted in a controlled and traceable manner in order to optimise the gain from measurement to measurement as background light conditions change.

The values P_0 and $P(r)$ found using Eqs. (5.10) and (5.11), respectively, can now be substituted into Eq. (5.9) and the calibrated attenuated backscatter calculated. However, as it is quite difficult to fully characterise the gain and efficiency of the system, and to monitor all of these in the field, an alternative method that can be used to calibrate attenuated backscatter using only measured signals is highly desirable.

5.3.2 Calibration Using Lidar Returns from Stratocumulus Layer

In the case of the prototype instrument used for preliminary field-testing at Chilbolton Observatory (51.15°N, 1.44°W) in Hampshire, UK, the gain of the APD was not calibrated before deployment and the gain at a reference reverse bias voltage and temperature specified by the manufacturer was applied instead. Due to the nonlinear gain behaviour of APDs, a large uncertainty in calculated gain can result if the gain is not known precisely at a specific voltage and temperature. The attenuated backscatter calculated using the best available parameterisation of the system deviated by more than two orders of magnitude from that given by a calibrated Vaisala CT75k research ceilometer at the test site. Therefore, the method of O'Connor et al. [3] described in Chap. 2 was applied. This same method was applied by O'Connor et al. in 2004 to calibrate the Chilbolton Observatory's CT75k.

As discussed in Chap. 2, this calibration method relies on the presence of a stratocumulus layer with a backscatter peak value of $\beta \geq 1 \times 10^{-4}\,\text{m}^{-1}\text{sr}^{-1}$ and the signal at this peak must be at least 20 times greater than the value 300 m higher. In addition, the calibration must be performed in the absence of drizzle, rain, and strong aerosol events. A stable stratocumulus layer at a height of about 750 m was observed between 15:00 and 16:00 UTC at Chilbolton in the afternoon on 4 September, 2012. A photograph of the cloud layer at 15:00 UTC is shown in Fig. 5.7.

Both the prototype and the CT75k were calibrated using lidar returns from this stratocumulus layer. Recall from Chap. 2, Eq. (2.39), that

Fig. 5.7 Image gathered by Chilbolton Observatory camera-axis 2100 at the beginning of calibration measurements at 15:00 UTC on 4 September 2012

$$\int_{0}^{\infty} \beta'(r) = \frac{1}{2\eta_m S}, \tag{5.12}$$

that is, the integrated attenuated backscatter return from a fully-attenuating cloud is equal to the reciprocal of twice the product of η_m, the multiple scattering correction factor, and S, the extinction to backscatter ratio. In order to perform the calibration, the factor $\eta_m S$ was determined for each instrument using a calculation table created by O'Connor [4]. The factors were calculated for a wavelength of 905 nm and a range of 750 m above the instruments, with the instruments considered to be located at sea level. The laser divergence and field of view of each of the instruments was used to calculate the correction factors. The prototype laser divergence is 0.35 mrad half-angle, with a half-angle field of view of 0.75 mrad. For the CT75k the half-angle laser divergence is 0.75 mrad and the half-angle receiver FOV is 0.6 mrad.

Figure 5.8 shows the attenuated backscatter measurements averaged for 1h for both instruments and calibrated using the principle of Eq. (5.12) by adjusting the scaling of the attenuated backscatter $\beta'(r)$ until the equation held true. The traces for each are shown with thin lines above and below designating the upper and lower bounds, respectively, of the calibration due to uncertainty in ηS as determined from O'Connor's table [4]. The calibrations of the two instruments do not quite fall within each other's uncertainty. Differences in range resolution between the two instruments (5 m for the prototype and 30 m for the CT75k) could lead the shapes of the traces to appear slightly different from each other. Also, since it is the integrated backscatter that is being calibrated to determine the correction factor, slight differences in profile shape between the two instruments will of course scale differently. As they are, the peak amplitudes of calibrated returns from the two instruments differ by around 15 %. Note that the correction multiplier found for the CT75k through the this calibration

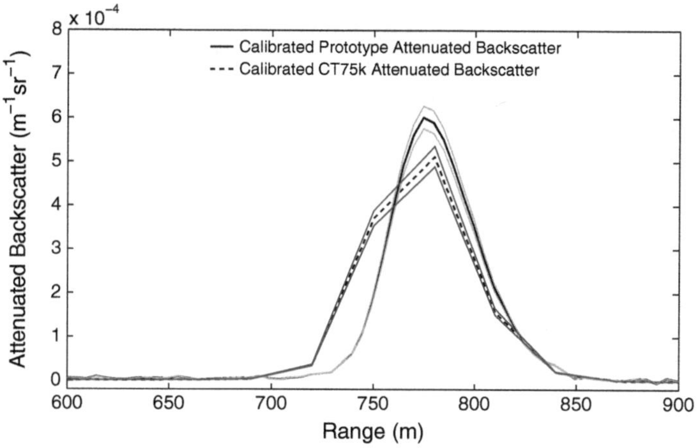

Fig. 5.8 Calibrated measurements of stratocumulus attenuated backscatter from the prototype and the CT75k at Chilbolton Observatory based on returns averaged for 60 min from 15:00 to 16:00 UTC on 4 September 2012

was a factor of 1.59 greater than that currently applied in the instrument settings at the observatory, which suggests a possible drift in the instrument, though only one calibration point was used for the current calibration.

5.3.3 Attenuated Backscatter Profile Examples

After application of the calibration method described by O'Connor et al., attenuated backscatter profiles measured by the prototype at Chilbolton were compared with those of the Vaisala CT75k research ceilometer (with its new calibration) located on the same site at three different times.

Figure 5.9 shows night-time measurements of boundary layer aerosols. The attenuated backscatter levels in this case were two orders of magnitude lower than those present during the calibration. Here the attenuated backscatter measured by the prototype, smoothed to match the 30 m range resolution of the other instrument, was a factor of 1.5 times greater than that measured by the CT75k. While there is this small difference between the amplitudes reported by the two instruments, it is clear from this example that the prototype instrument is capable of monitoring boundary layer aerosols. Mixed layer height can be determined from a lidar signal by, for example, finding the first significant negative gradient in the attenuated backscatter return [5]. By this definition the prototype would report the mixed layer height at 660 m while the CT75k would report it at 630 m for the measurement in Fig. 5.9.

Figure 5.10 shows attenuated backscatter measured from an ice cloud layer just below 8,000 m. Here again, the prototype output was smoothed to give it a 30 m range resolution. As in the previous case, the prototype measured a slightly larger attenuated backscatter level. Note that the prototype shows considerably more noise

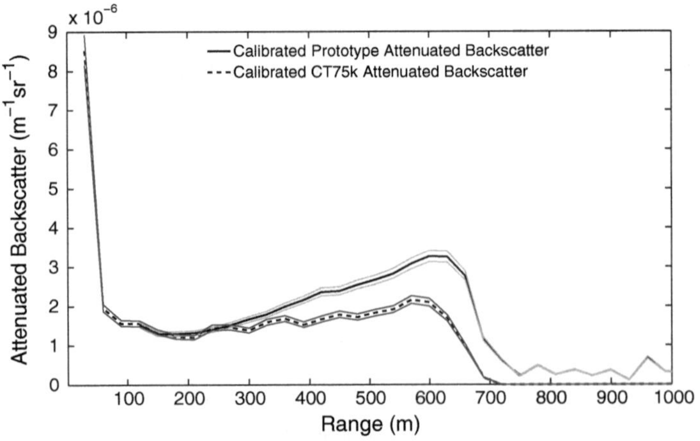

Fig. 5.9 Attenuated backscatter measurements from the prototype instrument and the CT75k at Chilbolton Observatory averaged for 30 min from 00:00 to 00:30 UTC on 2 September 2012

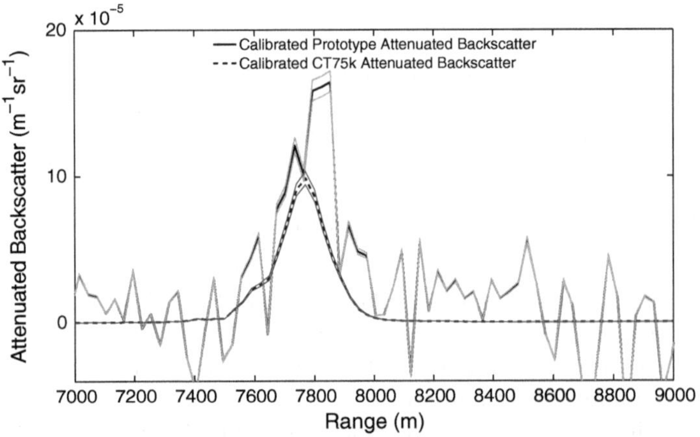

Fig. 5.10 Attenuated backscatter measurements from the prototype instrument and the CT75k at Chilbolton Observatory averaged for 60 min from 01:00 to 02:00 UTC on 4 September 2012

at this high range than the other instrument for two reasons. First, for the field testing of the instrument, it was configured to measure for only 2 s every 30 or 60 s for cloud height measurement. During the measurements shown in Fig. 5.10, the prototype was measuring for 2 s out of every 60, while the CT75k performed continuous averaging. Secondly, no noise suppression has been applied to the prototype output, while noise suppression is applied to the CT75k automatically.

A third and final example of attenuated backscatter measurements is shown in Fig. 5.11. For this low cloud measurement, the attenuated backscatter measured by the CT75k was, at the peak of the return, a factor of 1.3 greater than that measured by the prototype instrument. In the other comparisons, the prototype attenuated backscatter measurements were slightly greater than those measured by the CT75k. As multiple

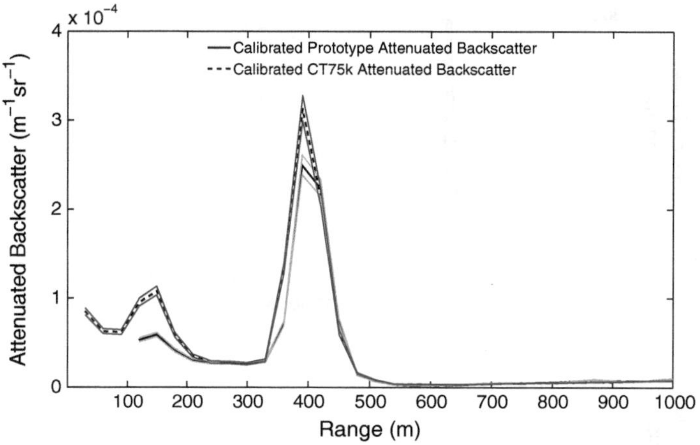

Fig. 5.11 Attenuated backscatter measurements from the prototype instrument and the CT75k at Chilbolton Observatory averaged for 30 min from 09:30 to 10:00 UTC on 2 September 2012

scattering effects from aerosols (Fig. 5.9) and ice clouds (Fig. 5.10) are considerably lower than those from water clouds, it would be expected that ratios for attenuated backscatter between the two instruments would be similar to the ratio between the two calibrations. However, it would also be expected that two instruments of different laser divergence and receiver field of view would show different values of attenuated backscatter for water clouds at heights other than the calibration height. For this reason O'Connor recommends that for a finely-tuned calibration, stratocumulus clouds from a variety of ranges should be used to derive the scalar correction for the instrument's integrated backscatter at each range, and then the scalar factor selected that minimises the error of the curve (Personal Communication, Ewan O'Connor 2013).

These three examples of different scattering media at different ranges demonstrate that calibration by the stratocumulus method at one range does not provide calibration for all instrument configurations at all ranges. This is not surprising considering the variability of multiple scattering effects in water clouds at different ranges and the effects of this variability on measured returns, as well as the possibility of unknown misalignments of the instruments or small unknown sensitivity variations in either instrument. The calibration does, however, provide a good reference point and the instruments in these varied examples showed a maximum ratio of one output to the other of about 1.5, a ratio which could have been orders of magnitude greater had the calibration not been performed.

5.4 Cloud Base Definitions

Figure 5.12 shows typical lidar returns from three different cloud types: a stratocumulus layer, an altocumulus layer, and a dense cirrus layer, all as measured by Platt et al. [6]. These three examples illustrate the fact that lidar return signals from clouds

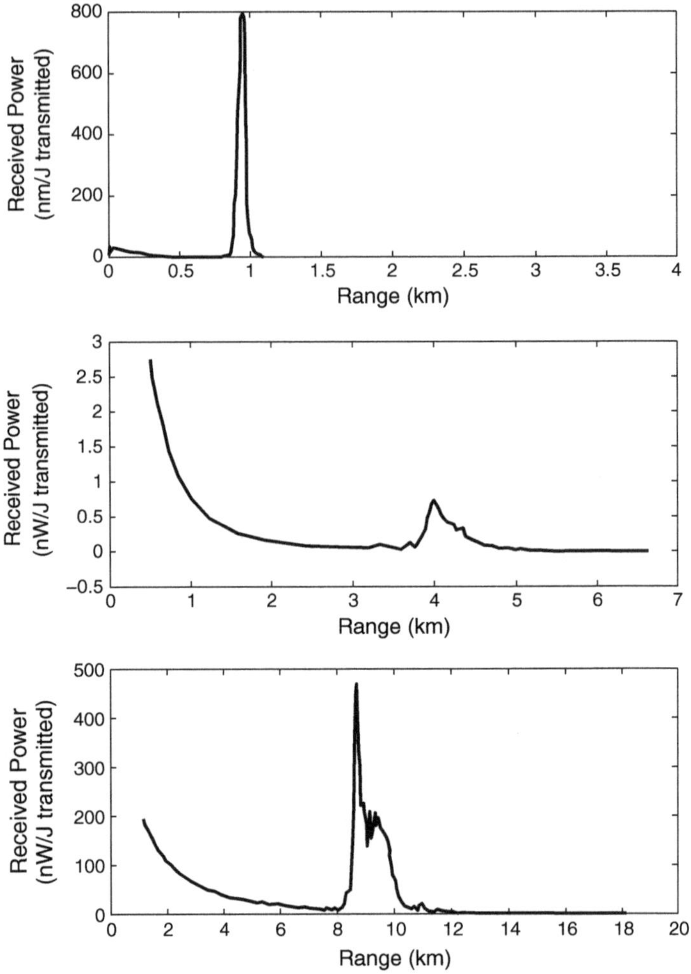

Fig. 5.12 Lidar returns from a stratocumulus layer (*top*), an altocumulus layer (*centre*), and a dense cirrus layer (*bottom*) recorded by Platt et al. [6]

vary considerably in width, amplitude, and shape. It is also clear from these examples that there are a variety of different possible ways to define cloud base in terms of the signal. In the altocumulus return, for example, it could be assigned at the peak at 4 km, at the onset at 3.5 km, or somewhere else. A useful automated cloud base detection algorithm must be based on definitions that can accommodate different types of cloud returns while also taking the effects of noise into account.

The most appropriate definition of cloud base may vary depending upon the application. For ceilometers, accuracy for aviation applications is a primary concern, so tying the cloud base definition to human perception of visibility is a top priority for the detection algorithm. This means, for example, that since extinction varies with

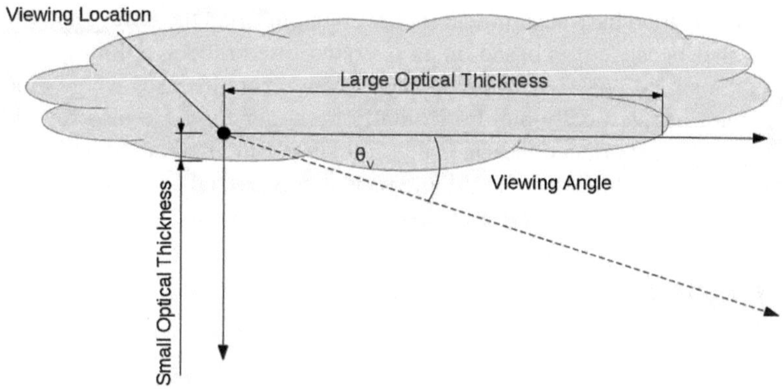

Fig. 5.13 Dependence of perceived optical thickness on viewing angle in a cloud

wavelength, if the laser wavelength strays too far from the visible region a wavelength correction may need to be applied. Furthermore, visibility near the base of a cloud varies greatly with viewing angle due to the difference in viewed cloud thickness as shown in Fig. 5.13. In this example, as the angle θ_V below horizontal increases, the optical thickness between the observer and the clear atmosphere outside the cloud decreases significantly.

In the following 10 sections, key methods from the literature for the determination of cloud base are reviewed, and the usefulness of each is considered particularly from the standpoint of ceilometer applications.

5.4.1 Peak Detection

The peak detection method searches a return signal from a cloud for the peak intensity at the top of an upward sloping cloud signal and assigns the cloud base to that height. Since it searches for the point of highest signal, this method is the least sensitive to the effects of noise. It is also the simplest method and the primary method used for visible spectrum searchlight-based rotating beam ceilometer (RBC) measurements. According to Eberhard [7], the peak of the measured signal was used for many years by the US National Weather Service as the definition of cloud base height.

In a study comparing ground-based measurements with airborne visual observations, Eberhard found that the peaks of the reported signals for lidar and RBC measurements agree well with each other and with nadir (downward vertical view) reports of pilots. However, Eberhard also found, from measurements of non-precipitating clouds at heights ranging from 160 to 3,200 m, that the peaks of the ground-based signals were on average 79 m above cloud base reports of pilots viewing at a typical landing approach angle of 3 degrees below horizontal. Thus the pilots lost visual contact with the ground well below the height of the signal peak. This corroborated earlier work by Eggert [8] which also showed that pilots viewing at a 3 degree slant angle reported cloud base at significantly lower heights than the RBC peak. Eberhard explained that because it is based on an oversimplified model, defining cloud base height as the signal peak will generate significant errors in cases where clouds are diffuse or are not vertically and horizontally homogeneous. He also reported that traffic controller experience indicates that if a simple method is required, the onset of signal agrees much better with pilot reports than does the peak of the signal.

5.4.2 First Derivative Zero Crossing

One way to locate the onset of a cloud return is to look for positive changes in the slope of the signal. Pal et al. [9] developed a cloud base height algorithm that monitors the first derivative of the signal in order to find the lowest significant returns from a cloud. The algorithm is based on the following reasoning. Once full overlap is reached, the returned signal in a clear atmosphere should decrease approximately exponentially. The presence of cloud particles should lead to an increase in the amount of backscattered signal detected, or at least a less than exponential decrease. The base of the cloud can be defined as the point where the first derivative of the signal crosses zero to become positive. Though it relies on a somewhat arbitrary definition, one advantage of this algorithm is that it can assign cloud base for a variety of types and densities of clouds in a consistent manner for meteorological study.

Figure 5.14, shows the altocumulus measurement from Fig. 5.12, this time along with its derivative. As is evident from this example, when there is noise or a varying return from thin layers near the cloud base, the onset of signal is harder to fix precisely than the peak position. Three points near the cloud base (at 3.3, 3.7 and 3.8 km) show positive derivatives immediately following non-positive derivatives. In this case confusion could be avoided simply by testing for two or three successive positive slope bins starting from the first positive slope bin in order to filter out spurious results; in that case the cloud base would be assigned at 3.8 km. Based on this example, it is clear why Pal et al. noted that this detection method does not work well for signals

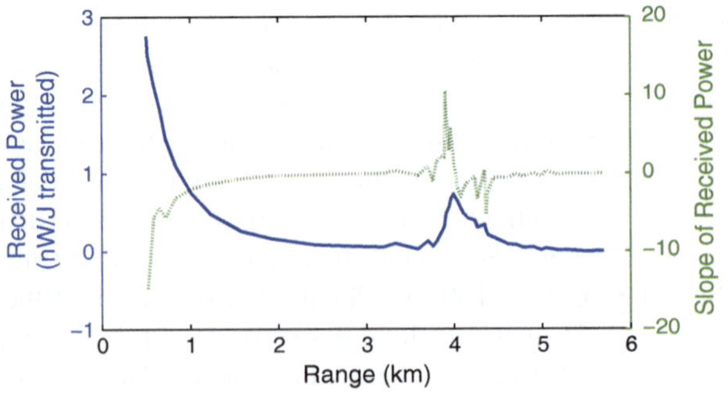

Fig. 5.14 Altocumulus lidar return (*blue*) from Fig. 5.12 and its derivative (*dashed green*)

that do not show a monotonic decrease in clear air, for signals in fog or precipitation conditions, or for signals from clouds that do not show a clear signal increase at the cloud base. A relatively noisy lidar signal such as that from a ceilometer would also be expected to be difficult to process with this method.

Besides setting a minimum number of consecutive range bins of positive slope in order to flag a cloud, these problems could be partially addressed by changing the threshold value for the derivative. For instance, it could be set to some small positive value for the algorithm to work in noisy conditions, or it could be set to a small negative value for cases of low density cloud in which returns from the clouds are not substantial enough to fully overcome the exponential decrease of the signal.

5.4.3 Method of Clothiaux et al.

Clothiaux et al. [10] improved the slope-based cloud detection method somewhat by comparing the slopes of clear sky returns with those of cloudy sky returns in order to locate cloud edges. Before individual signals are evaluated, some pre-processing is required. First, an experimentally-derived threshold is used to distinguish periods of clear sky from periods of cloudy sky in the data. Then recent clear sky period signals are averaged to determine the current clear air return. The molecular extinction and backscatter coefficients, $\alpha_{\mathrm{mol}}(r)$ and $\beta_{\mathrm{mol}}(r)$, respectively, are derived using pressure and temperature measurements of the local atmosphere and applied to calibrate the clear air lidar measurements and therefore account for changes in the lidar instrument pulse power, overlap, sensitivity, etc.

After pre-processing, the ratio of an individual measured cloud signal can be compared with the averaged clear air return as shown in Fig. 5.15. When the ratio is constant with range, a clear atmosphere is assumed, and when the ratio is changing with range, a cloud layer is assumed. In the example shown in Fig. 5.15, the dotted

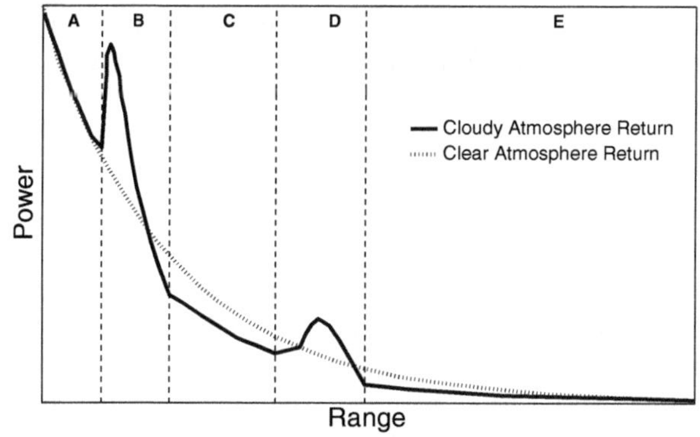

Fig. 5.15 Schematic example of signal zones designated by the method of Clothiaux et al. [10]

trace indicates a model clear air return and the solid line indicates a cloudy return. The ratio between these returns is roughly constant within regions A, C and E, whereas it is changing in regions B and D. This gives the general positions of the various cloud layers during the cloudy periods. To define cloud edges for these identified layers, the backscatter coefficient $\beta(r)$ is determined by inversion of the lidar equation, and the cloud base height is assigned at the altitude where the measured backscatter exceeds an experimentally-determined threshold.

This method can be effective for remote instruments in climates where the lidar return contour of the clear atmosphere can be measured, for example via radiosonde, or modelled precisely. Clothiaux et al. explained that their method has been used successfully on data from the Micropulse Lidar [11], a research-grade eye-safe elastic lidar. However, for an instrument such as a ceilometer with low sensitivity to molecular returns, this approach is unlikely to perform well, since many hours of averaging can be required to obtain a single clear sky return profile. In addition, this method is best suited for processing of data only after a significant number of measurements have been made, which is problematic for an instrument that is required to provide real-time outputs every 30 s. The definition of cloud base height based on a threshold on the backscatter coefficient, however, is similar in many ways to the visibility methods discussed later in this chapter, which provide the starting point for the algorithm applied in this thesis.

5.4.4 Method of Winker and Vaughan

Rather than relying on the measurements of the signal itself to identify clear air returns and then using a model to generate a backscatter threshold as Clothiaux et al. did, Winker and Vaughan [12] applied a threshold to the ratio of measured attenuated backscatter returns to calculated clear-atmosphere molecular returns. For noise-free measurements the threshold for this ratio could be set to unity. However, in practise a threshold, S_c, greater than unity is used.

Since noise generally increases with range on a range-corrected attenuated backscatter signal, for returns from farther ranges they applied a range-dependent threshold, $S_c(r)$, calculated by

$$S_c(r) = 1 + \frac{S_c(r_0)}{\beta'_{\text{mol}}(r)}, \tag{5.13}$$

where $S_c(r_0)$ is a threshold that depends on the noise level of the signal and $\beta'_{\text{mol}}(r)$ is the calculated molecular attenuated backscatter signal which decreases exponentially with range. This means that the threshold $S_c(r)$ will increase with range at a rate that depends on the $S_c(r_0)$ value selected. The range-variable threshold, $S_c(r)$, is designed for detection of higher clouds, and the constant threshold, S_c, is generally preferred for lower clouds since $\beta'_{\text{mol}}(r)$ may be so large that $S_c(r)$ approaches a value of 1. At any given range the greater of these two thresholds is applied.

In addition to these amplitude thresholds that account for the slope of the signal, width thresholds are applied. For most clouds, the amplitude threshold needs to be exceeded for seven consecutive range bins, where the variable range resolution is 6, 15, or 30 metres per bin. If this criterion is not met, a second width test for thin clouds is applied. In this test a feature above threshold for three consecutive range bins in which at least one range bin exceeds a higher threshold, S_t, will be identified as a thin cloud layer.

Since the threshold variables S_c, $S(r_0)$, and S_t are not tied precisely to physical definitions, they require experimental fine-tuning. This is not unusual for cloud detection algorithms, however, and this method has been further developed for use on space-based lidar signals [13, 14]. For lidar ceilometry the fact that the sensitivity to molecular returns is very small would again make this method difficult to apply. However, the general techniques of range-variable amplitude thresholding and variable width thresholding are useful techniques, and different width thresholds for wide and narrow clouds are applied in the current work.

5.4.5 Method of Platt et al.

A related method described by Platt et al. [6] looks at the difference between the received signal and a stored clear background measurement for each range bin. Two criteria must be met in order to assign a cloud base. First, the amplitude of this difference must exceed a selected multiple of the standard deviation of the signal noise. Second, this amplitude must be maintained or exceeded for a specified minimum number of range bins. The cloud base is then assigned to the first bin above threshold.

The amplitude difference threshold filters the low level noise, while the duration (width) requirement helps remove spurious noise spikes. As with the slope method, this threshold method may have difficulty detecting a diffuse cloud, as the difference between the cloud signal and the background signal may not be much more than the standard deviation of the noise. However, setting the threshold low enough to detect thin clouds makes the instrument more susceptible to noise. This trade-off can be optimised through tuning of the thresholds, but the tuning would be expected to vary somewhat depending upon the atmospheric conditions during the measurement.

5.4.6 Method of Campbell et al.

Campbell et al. [15] proposed a way reduce the noise-susceptibility problems of threshold methods like the previous one by using an approach that sets two different thresholds that must both be exceeded.

First, the normalised relative backscatter (NRB), which is equal to the attenuated backscatter, β', times a dimensional system calibration constant, $C_{\beta'}$, such that

$$\mathrm{NRB(r)} = C_{\beta'}\beta'(r), \tag{5.14}$$

must show an increase of 55 % over one or two range bins.

Second, the signal to noise ratio must show an increase of approximately 42 % over the same one or two range bins. The SNR can be expressed (in a form simplified from that expressed by Campbell et al.) as

$$\mathrm{SNR}(r) = \frac{\frac{\mathrm{NRB}(r)}{r^2}}{\sqrt{\frac{\mathrm{NRB}(r)}{r^2} + N_B(r)}}, \tag{5.15}$$

where $N_B(r)$ is the background noise received during the measurement from each range r. Note that since the NRB is a range-corrected function, here it is divided by r^2 in order for it to be expressed as part of the signal to noise ratio.

Equation (5.15) demonstrates the noise-dependent relationship between SNR and NRB. In low background light conditions the NRB is the more restrictive threshold, since in that case the SNR will be larger than required. In bright background conditions the SNR is the more restrictive threshold, since in that case the NRB will be larger than required. This dual threshold approach provides an attractive means of dealing with differences between daytime signals and nighttime signals.

When applied to ceilometers, however, the problem with any thresholding approach is the relatively large amount of noise. Since wide laser spectra and therefore wide bandpass filters are typically used in ceilometers, as are fairly wide fields of view, the noise due to background light is much greater than that present in most research lidar systems. Note that in contrast to the prototype instrument described in this thesis which has a filter bandwidth of 36 nm, the Micropulse Lidar to which this method was applied has a filter bandwidth of 0.2 nm [11]. This means approximately 180 times more optical background noise for the ceilometer during daytime measurements based on the filter bandwidth alone. In addition, while the prototype has a half-angle field of view of 0.75 mrad, the Micropulse Lidar has a half-angle field of view of 0.05 mrad, which means an additional factor of 225 times more noise for the ceilometer during daytime measurements. Since the signal to noise ratio of a ceilometer is much smaller than that of a high performance instrument like the Micropulse Lidar, it is also more likely that spurious 55 % increases in NRB over one or two range bins will be present in the ceilometer signal.

5.4.7 Method of Gaumet et al.

Gaumet et al. [16] described a cloud detection method specifically for ceilometers based on first identifying cloud signal onset, peak, and top, and then applying a threshold test. These features are located, either on the signal itself or on the inversion of the signal, by performing sliding derivative tests on three consecutive range bins. The signal onset is identified at the first significant positive derivative, the peak is identified just before the following significant negative derivative, and the top of the cloud is located when the negative slope concludes just before a roughly constant slope is reached.

If the signal itself is used, a threshold on the amplitude of the peak of the signal is applied to determine whether the cloud is a significant layer or not. It the inversion is used, the optical depth of the layer is calculated and compared with a threshold to determine its significance. These thresholds were established through comparison with visual observation. If the layer is considered significant, the height of the peak of the signal within the layer is assigned as the cloud base height. This type of method should perform well at identifying visible cloud layers in situations without excessive noise, though it would be better for aviation applications to locate the cloud base at the onset of the signal. In the presence of considerable noise, however, this derviative-based method would be expected to struggle.

5.4.8 Structure of the Atmosphere (STRAT) Method

Morille et al. [17] described a comprehensive method for identifying aerosol and cloud layers throughout the troposphere from lidar signals. The method, called Structure of the Atmosphere (STRAT) requires a signal to noise ratio of at least 3 in order for any layers to be considered. If the signal meets this criterion, layers containing aerosol or cloud droplet particles are detected using a continuous wavelet transform method based on the second derivative a of Gaussian distribution, known as the *Mexican hat wavelet*. This particular wavelet, shown in Fig. 5.16, is used because it closely resembles the lidar signature of a cloud or aerosol layer.

If a feature is identified through wavelet analysis, it must also exceed a peak to base ratio threshold, R_{thr}, in the signal $P(r)$ such that

$$R_{\text{thr}} = \frac{P(r_{\text{peak}})}{P(r_{\text{base}})} = 10\sigma(t), \tag{5.16}$$

where $P(r_{\text{peak}})$ is the calibrated signal power at the peak of the feature identified by the wavelet analysis, $P(r_{\text{base}})$ is the calibrated signal power at the onset of the

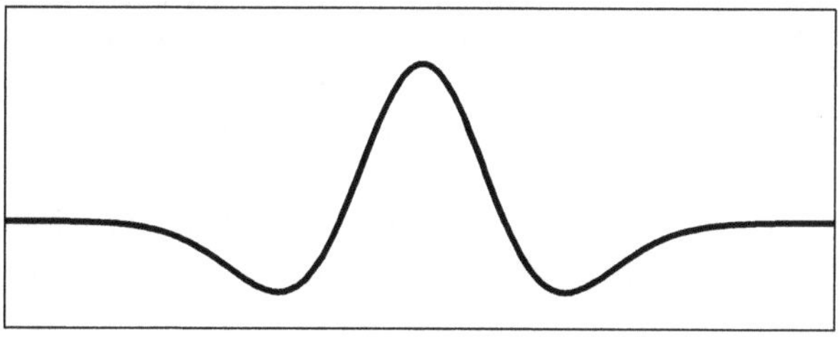

Fig. 5.16 Second derivative of a Gaussian distribution, referred to as the *Mexican hat wavelet*

feature identified by the wavelet analysis, and $\sigma(t)$ is the standard deviation of the calibrated lidar signal $P(r)$ at time t.

Once a significant particle layer has been identified in this way, cloud layers are distinguished from aerosols by application of another threshold. To do this a test, originally devised by Wang and Sassen [18], is applied in which the calibrated, range-corrected peak to base ratio δPr^2 is considered such that

$$\delta Pr^2 = \frac{P(r_{\text{peak}}, t) r_{\text{peak}}^2}{P(r_{\text{base}}, t) r_{\text{base}}^2}. \tag{5.17}$$

If $\delta Pr^2 > 4$, the layer is determined to be a cloud.

Morille et al. found that a co-located Vaisala ceilometer typically reported cloud base height between the onset and the peak of the signal. In the STRAT algorithm, cloud base is assigned at the onset of the lidar signal as determined by analysis of the slope, and therefore reported cloud base on average 178 m below the ceilometer.

This method is effective because it does not rely on inverted data, and because it combines threshold and slope analysis with feature detection. It is intended, however, as a comprehensive algorithm for vertical profiling of the atmosphere utilising lidar systems which show clear molecular returns, as the molecular returns are used in the calibration of the signal. While a wavelet approach was investigated for the prototype, it was not applied, as it was not found to show advantage over the method that was eventually implemented. However, the concept of signal filtering for feature identification as a separate step from thresholding was applied to the prototype.

5.4.9 Fraction of Total Signal (FOTS) Method

In the same study in which he suggested that from a visibility perspective lidar signal onset was a more accurate definition for cloud base than signal peak [7], Eberhard went on to suggest a more complex approach called the fraction of total signal (FOTS) method to relate lidar signals to pilot-reported cloud base height at various viewing angles in a variety of non-uniform cloud density distributions.

As illustrated previously in Fig. 5.13, the optical thickness perceived by a pilot near the bottom of a cloud varies significantly with viewing angle. In order to calculate the height, r_e, at which a pilot will report cloud base for a particular viewing angle, θ_V, below horizontal, based on a lidar return, Eberhard applied two assumptions. First, he assumed that the lidar ratio $\Pi_\rho(r)$ defined in Eq. (2.14) is constant with range such that $\Pi_\rho(r) = \Pi_\rho$. Then, since the particles in clouds are relatively large, he assumed a fixed ratio, ρ, of extinction, α_L, at the lidar wavelength to extinction, α_P, observed by the pilot such that $\rho = \frac{\alpha_L}{\alpha_P} = \text{constant}$. He also fixed a standard contrast threshold of 0.05 of perceived contrast C_D over actual contrast C_0 at distance D such that

$$\frac{C_D}{C_0} = e^{-\int_0^D \alpha(s) ds} = 0.05. \tag{5.18}$$

By integrating extinction up to distance D, the optical depth at this contrast ratio is found by

$$\tau_e = \int_0^D \alpha(s)ds = 3. \tag{5.19}$$

Note that contrast can be defined as the difference in relative brightness between an object and its surroundings [19].

Using the assumptions of constant lidar ratio and constant ratio ρ, Eberhard derived an expression, slightly rewritten here, that can be used to find cloud base height perceived by a pilot from a ground-based lidar return. For a vertically-pointing lidar, when the ratio of attenuated backscatter β' integrated up to range r_e to the total integrated attenuated backscatter in a cloud meets the following criteria,

$$\frac{\int_0^{r_e} \beta'(r)dr}{\int_0^{\infty} \beta'(r)dr} = 1 - e^{-2\rho \sin(\theta_V)\tau_e}, \tag{5.20}$$

where the viewing angle θ_V and the optical depth threshold τ_e are user selected, the height r_e is reported as the cloud base height.

Dependence on the unknown lidar ratio Π_ρ and the unknown system constant K_s, both accounted for in the attenuated backscatter $\beta'(r)$, is therefore removed when the two integrals are divided. In order to remove the dependence on infinite distance for the total integrated attenuated backscatter, Eberhard assumed that any cloud under measurement is sufficiently optically dense to fully attenuate the signal before the top of the cloud is reached. He noted, however, that for thin clouds that do not fully attenuate the signal, the FOTS method can still assign cloud base height at least as accurately as can the peak detection method.

To explore how this method performs it was applied, as demonstrated in Figs. 5.17 and 5.18, to two example measurements of Platt et al. shown earlier in Fig. 5.12. Note that neither of these returns is range-corrected, and that the signal was integrated starting from just below each cloud. Though the example in Fig. 5.18 shows a much broader return than that in Fig. 5.17 and is presumably considerably less dense, based on the approximately exponential shape of the decay of the cloud return in each example it seems reasonable to assume that both fully attenuated the signal.

The perceived cloud base was calculated from the lidar return for each of these examples for pilot viewing angles, θ_V, equal to 1°, 3° and 90° below horizontal. Note that the perceived cloud base height calculated using a viewing angle of 90° below horizontal occurs near the height of full extinction of the received lidar signal, while for the smaller viewing angles cloud base height is reported below the peak of the return. The 3° report is approximately 100 m below the 90° report for the stratocumulus example in Fig. 5.17 and more than 1 km below it in the altocumulus example in Fig. 5.18. Perception of cloud base height is therefore highly dependent

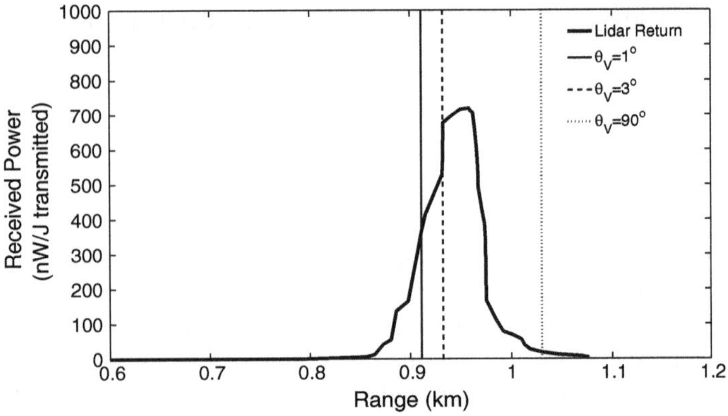

Fig. 5.17 Example of cloud base location by Eberhard's FOTS method applied to the stratocumulus lidar return shown in Fig. 5.12. Cloud base heights for different viewing angles are indicated by *vertical lines*

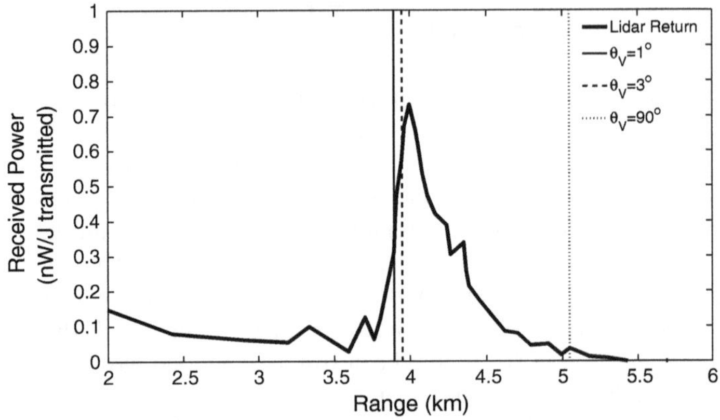

Fig. 5.18 Example of cloud base location by Eberhard's FOTS method applied to the altocumulus lidar return shown in Fig. 5.12. Cloud base heights for different viewing angles are indicated by *vertical lines*

upon viewing angle, and this angle should always be specified for cloud base definitions based on visibility.

The FOTS method provides an interesting way of linking cloud base height to optical properties without relying on assumptions necessary for inversion of the lidar equation. The assumption of lidar signals being fully attenuated within clouds is not a bad one, since lidar signals rarely penetrate substantial clouds. However, when this is not the case, the definition used in this approach loses its meaning. The following method is also based on visibility, but the assumptions necessary for Klett inversion are applied, rather than the signal attenuation assumption applied by Eberhard.

5.4.10 Method of Poyer and Lewis

Poyer and Lewis [20] reported a cloud height method based on horizontal visibility determined from a Klett inversion of the atmospheric extinction coefficient, $\alpha(r)$. Since visibility can be determined from extinction using Koschmieder's law for visual meteorological optical range V_{MOR} [21],

$$V_{\mathrm{MOR}} = \frac{\ln K'}{\alpha}, \tag{5.21}$$

where K' is the luminance contrast threshold of the human eye, a horizontal visibility at each height can be calculated from the inverted signal. Though there are differences in perceived contrast at daytime and nighttime [22], the typical visual contrast threshold as dictated by the World Meteorological Organization [23] and the International Civil Aviation Organization [24] is 5 %. Applying this threshold and assuming horizontal homogeneity at each height, Poyer and Lewis calculate horizontal visibility $V_H(r)$ as a function of height r from the retrieved extinction profile $\alpha(r)$ by [20]

$$V_H(r) \approx \frac{3}{\alpha(r)}. \tag{5.22}$$

Poyer and Lewis then applied empirically-derived thresholds for horizontal visibility and slope of horizontal visibility to determine cloud base height from Micropulse Lidar signals. Because this approach ties the cloud base reported to visibility, it makes it appropriate for use in aviation, but as it relies on clearly defined optical properties of the cloud, it is also relevant for meteorological use. This method forms the basis for the first part of the algorithm applied to the prototype data and is described in more detail in Sect. 5.5.

5.4.11 Summary of Cloud Base Determination Techniques

Each of these methods has advantages and disadvantages. Peak detection is the simplest method but typically overestimates the height at which a pilot can see out of a cloud, and as Eberhard pointed out, there does not seem to be a simple formula that reliably relates cloud base height to signal peak [7]. The slope and threshold methods have advantages for research lidars, particularly the threshold method of Clothiaux in the setting of instruments where current atmospheric state data is available. Eberhard's FOTS method considers pilot perception of cloud base as its basis and is therefore relevant to ceilometers, but its reliance on the assumption of optically dense clouds means it may significantly under-report the heights of thin cloud bases. The method of Poyer and Lewis, which was used as the starting point for the prototype algorithm, ties the signal to visibility in a straightforward manner.

5.5 Prototype Ceilometer Cloud Base Algorithm Part 1: Visibility-Based Detection

In order for a lidar ceilometer to report cloud height automatically, an algorithm must be applied to the data inside the instrument. This algorithm must not only determine whether or not cloud detections are significant and assign cloud base height accurately, but it must also avoid false reports in precipitation, discern whether apparently independent cloud detections within close proximity to each other are actually part of the same layer, and create outputs that can be used by standard meteorological reporting frameworks. This section details the first part of the two-part cloud base algorithm that has been developed for the prototype instrument.

5.5.1 Horizontal Visibility Thresholds for Two Cloud Types

The first part of the algorithm that has been developed is based on the visibility-based method of Poyer and Lewis [20]. By applying Eq. (5.22), horizontal visibility is determined for each height of the inverted signal.

In order to detect cloud base height from the lidar returns, Poyer and Lewis applied visibility threshold rules that were derived empirically by comparing ground-based visual observations of cloud base height to data inverted by the Klett method. Their algorithm works as follows.

For dense water droplet clouds, the following criteria are applied to identify cloud base.

$$\text{If } V_H(R) < V_d \text{ and } [V_H(R+1) - V_H(R-1)] > \Delta V_d, \text{ then } R = R_c. \quad (5.23)$$

Here $V_H(R)$ is the horizontal visibility at range bin R, calculated using Eq. (5.2), V_d is the visibility threshold for dense cloud base, $V_H(R+1)$ and $V_H(R-1)$ are the visibilities calculated 1 range bin higher and 1 range bin lower, respectively, than range bin R, ΔV_d is the visibility gradient threshold, and R_c indicates a cloud base.

The threshold for dense cloud base chosen by Poyer and Lewis,

$$V_d = 0.64\,\text{km} \quad \text{(Poyer and Lewis)}, \quad (5.24)$$

sets the horizontal visibility (as expressed by Eq. 5.21) that an observer would perceive when standing at the base of a cloud to match the visibility that would be observed when standing on the ground in a fairly dense fog (Personal Communication, Aaron Poyer 2009). Meteorological expertise within Campbell Scientific suggests that this threshold should be raised slightly, therefore lowering the height at which the base of a cloud would be reported. A threshold of $V_d = 1\,\text{km}$ was suggested, since that is the visibility limit at which fog is first reported [25], rather than demanding the visibility decrease to that of a dense fog before a cloud can be reported. This adjusted threshold is intended to match the cloud base reports of the instrument more closely with those reported by pilots. Therefore, for the algorithm used in the prototype, the threshold

$$V_d = 1\,\text{km} \quad \text{(Prototype)} \tag{5.25}$$

is applied for dense cloud base detection.

The range bin size of the Micropulse Lidar instrument used by Poyer and Lewis was 15 m. The gradient threshold they applied for this range bin was

$$\Delta V_d = 0.064\,\text{km} \quad \text{(Poyer and Lewis).} \tag{5.26}$$

Since the prototype ceilometer has 5 m range bins, the gradient threshold

$$\Delta V_d = \frac{0.064\,\text{km}}{3} = 0.022\,\text{km} \quad \text{(Prototype)} \tag{5.27}$$

was used to give an equivalent definition.

For thin water droplet and/or ice crystal clouds that do not meet the criteria in Eq. (5.23), Poyer and Lewis applied a different rule.

$$\text{If } \frac{\sum_{R}^{R+20} V_i}{20} < V_{\text{ave}} \text{ then } R = R_c. \quad \text{(Poyer and Lewis)} \tag{5.28}$$

Here V_i is the horizontal visibility calculated for range bin i, and V_{ave} is the average visibility over 20 of the micropulse lidar's 15 m range bins, or 300 m. The average visibility threshold specified by Poyer and Lewis is

$$V_{\text{ave}} = 4.8\,\text{km} \tag{5.29}$$

In the prototype algorithm, the same criterion requiring a 300 m vertical extent with average horizontal visibility below V_{ave} was applied in order to indicate a thin cloud, but the cloud base height was assigned at the centre height of the 300 m region rather than at the bottom, placing the cloud base 150 m above the lowest bin exceeding the threshold in an attempt to agree more closely with pilot observations of visibility. In order to centre the cloud base height assigned to a thin cloud feature and employ 5 m range bins rather than 15 m range bins, the rule from Eq. (5.28) becomes the following.

$$\text{If } \frac{\sum_{R-30}^{R+30} V_i}{61} < V_{\text{ave}} \text{ then } R = R_c. \quad \text{(Prototype)} \tag{5.30}$$

This rule was used on the prototype data to locate thin cloud layers.

5.5.2 Inversion Boundary Point Assignment, and Minimum Measurable Extinction Coefficient

After the data has been corrected for the impulse response of the electronics, overlap corrected, and range corrected, it is inverted using Klett's backward inversion method. Recall from Chap. 2 that this can be expressed as

$$\alpha(r) = \frac{Z(r)}{\frac{Z(r_b)}{\alpha(r_b)} + 2\int_r^{r_b} Z(r')dr'}, \tag{5.31}$$

where $Z(r)$ is the range-corrected (here also overlap-corrected) power as a function of range, $Z(r_b)$ is the range-corrected power at the far boundary r_b (the far boundary being the maximum range at which the signal to noise ratio is deemed to be acceptable), $\alpha(r_b)$ is the assumed far boundary value of the extinction coefficient, and $\int_r^{r_b} Z(r')dr'$ is the range-corrected power integrated from range r to range r_b.

Typically, the starting range of the inversion is the range at which full overlap has been reached, however, since the overlap function of the instrument has been characterised, inversion of the prototype data begins at 80 m, the height of onset of overlap. This is the first range from which single-scattering returns can be detected. Empirically-derived visibilities for strong multiple scattering returns are used below this height.

The boundary range r_b is determined by finding the farthest point where the signal to noise ratio is 2 times the noise found at the maximum range of the instrument and is immediately preceded by at least three consecutive range bins also exceeding this threshold. The noise is found by taking the standard deviation of the top 500 m of the measurement range where very little, if any, backscatter signal is expected to be detectable.

In an elastic lidar application where the extinction coefficient at the far boundary is not known, it must be assumed. For turbid atmospheres where clouds are the primary features of interest, the Klett inversion typically converges quickly to the correct solution regardless of the assumed boundary extinction coefficient [26]. An experiment on data from the prototype showed that a change of 2 orders of magnitude in the boundary extinction coefficient $\alpha(r_b)$ produced a much smaller change of a factor of 2 in the inverted extinction coefficient at close ranges in a cloudy atmosphere.

Boundary values for $\alpha(r)$ reported in the literature range from approximately 0.02/m for cloudy boundaries [27] to 0.0001/m for clear boundaries [16]. Due to limited sensitivity to molecular returns (on the order of 8 h of averaging at night time is required to see a ceilometer molecular return [28]), it is assumed that structures detected at the far boundary are very likely to be either clouds of some sort or highly concentrated aerosol layers. Because of this, and also to make sure that the highest clouds are not missed by the visibility threshold, a fairly high boundary extinction value of $\alpha(r_b) = 0.01\text{m}^{-1}$ is used. If this guess is higher than the actual extinction coefficient at the boundary, greater values of $\alpha(r)$ at far ranges are calculated and therefore lower visibilities, making cloud features at far ranges more likely to show up. Rather than leave out high clouds that are actually there, the high estimate approach is preferred; this is balanced by the second part of the algorithm, which is designed to remove false cloud hits as discussed in the next section.

5.6 Prototype Ceilometer Cloud Base Algorithm Part 2: Bandpass Filtering and Thresholding

In order to filter out spurious cloud reports generated by the visibility threshold method, particularly at the far end of the measurement range where the inversion of the extinction coefficient profile is highly sensitive to noise, a second cloud detection method not requiring inversion was developed to run in parallel with the visibility-based algorithm. Only when both methods detect a cloud base at a given height is that cloud base considered significant and therefore reported.

Initially, a matched filtering approach was applied to the data in the time domain. Gaussian or Hanning (cosine) windows on various scales were convolved with measured data in order to enhance features of that shape in noisy situations. The problem with this method was that a broad non-cloud feature such as light haze or precipitation could push the filtered signal above threshold, and then any returns from smaller-scale clouds within that region would not show up since the threshold was already crossed by the returns from the broad feature. Rather than attempting to optimise a set of thresholds with variable amplitudes and feature widths, the following method was used.

5.6.1 Filter Specification

Two frequency domain bandpass filters, shown in Fig. 5.19, were selected to enhance signal features of widths corresponding to typical broad and narrow cloud returns. Bandpass filter 2, $B_2(\omega)$, was generated using low-frequency roll-on and high frequency roll-off from an inverted Gaussian profile. Using 1024 frequency bins ω_i (where ω_{1024} corresponds to a frequency of 15 MHz), bandpass filter 2 was generated such that

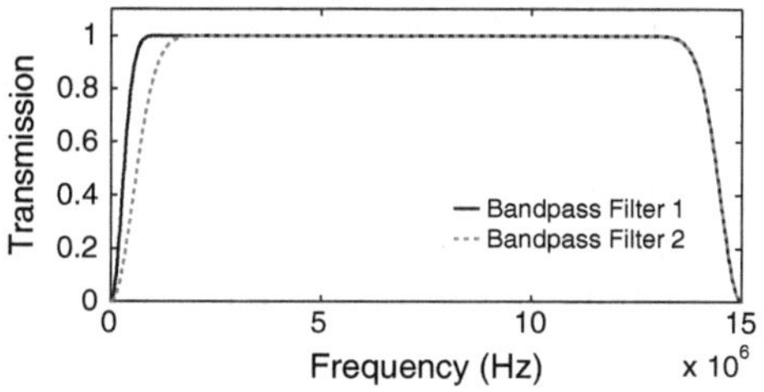

Fig. 5.19 Bandpass filters 1 and 2

$$B_2(\omega_i) = \begin{cases} 1 - e^{-(\frac{\omega_i}{50})^2} & \omega_i \leq 512 \\ 1 - e^{-(\frac{\omega_i - 1024}{50})^2} & 512 < \omega_i \leq 1024. \end{cases} \tag{5.32}$$

Bandpass filter 1, $B_1(\omega)$, has the same high frequency roll-off as filter 2, but it has a faster low frequency roll-on from a different Gaussian profile such that,

$$B_1(\omega_i) = \begin{cases} 1 - e^{-(\frac{\omega_i}{25})^2} & \omega_i \leq 512 \\ 1 - e^{-(\frac{\omega_i - 1024}{50})^2} & 512 < \omega_i \leq 1024. \end{cases} \tag{5.33}$$

These specific filters were selected because when applied to the signal they successfully identified the wide variety of cloud layer shapes that were manually identified in the test dataset.

Bandpass filter 1, $B_1(\omega)$, is applied to the Fourier transform of a time domain signal $P(t)$ that has been corrected for impulse response, but not corrected for overlap or range. The filtered signal in the time domain $P_{B_1}(t)$ is found by

$$P_{B_1}(t) = \mathcal{R}\Big[\mathcal{F}^{-1}\Big[B_1(\omega) \cdot \mathcal{F}[P(t)]\Big]\Big]. \tag{5.34}$$

The same method is applied using bandpass filter 2, $B_2(\omega)$ in order to find the second filtered signal, $P_{B_2}(t)$.

In order to identify potential cloud features from the filtered signals, the standard deviation is calculated for the top 500 m of the range, and a threshold is set to four times this value. Regions in the filtered signals that exceed this threshold are flagged as potential clouds. As with the shapes of the filters, this threshold was optimised by comparison with manual analysis of the test dataset.

As shown in Fig. 5.19, filter 1 and filter 2 share the same high frequency roll-off, however, filter 1 turns on faster than filter 2, reaching 50 % by 308 kHz, while filter 2 doesn't reach 50 % until 615 kHz. Filter 1 is therefore expected to enhance wider features in the time domain than filter 2. The performance of these filters and their application along with the visibility threshold method is demonstrated in the next section.

5.7 Parallel Algorithm Method, Cloud Layer Reporting Rules, and Cloud Detection Algorithm Results

In order for the algorithm to report a cloud base, two criteria must be met. First, a cloud feature must be identified by the bandpass filter method just discussed, and second, the inverted cloud return must pass either the dense cloud test or the thin cloud test in the horizontal visibility threshold method. Only when both of these methods locate a cloud will cloud base be reported. Two example measurements illustrate how these methods work together in order to determine if a significant cloud layer is present.

5.7.1 Cloud Detection Algorithm Examples

Figure 5.20 shows the application of the cloud detection algorithm to a cloud located at an altitude of approximately 3,500–4,000 m at Chilbolton Observatory early in the morning on 5 June, 2012. The upper graph shows the Impulse Response-corrected (IR-corrected) signal measured by the prototype. Potential cloud features identified using the bandpass filters are indicated by the shaded areas, dark grey for wider features identified using bandpass filter 1, and light grey for narrower features identified using bandpass filter 2. The lower graph shows the extinction profile, in green, derived from the IR-corrected signal by Klett inversion and the horizontal visibility, in blue, at each range calculated from the extinction profile. Here a point meeting the thin cloud visibility threshold criteria is indicated by a dashed vertical line and a point meeting the dense cloud visibility threshold is indicated by a solid vertical line.

In this example, potential close range features arising from noise spikes and/or aerosol returns detected by the bandpass method in the first 200 m of the upper graph are not determined to be clouds since range correction and inversion in the lower

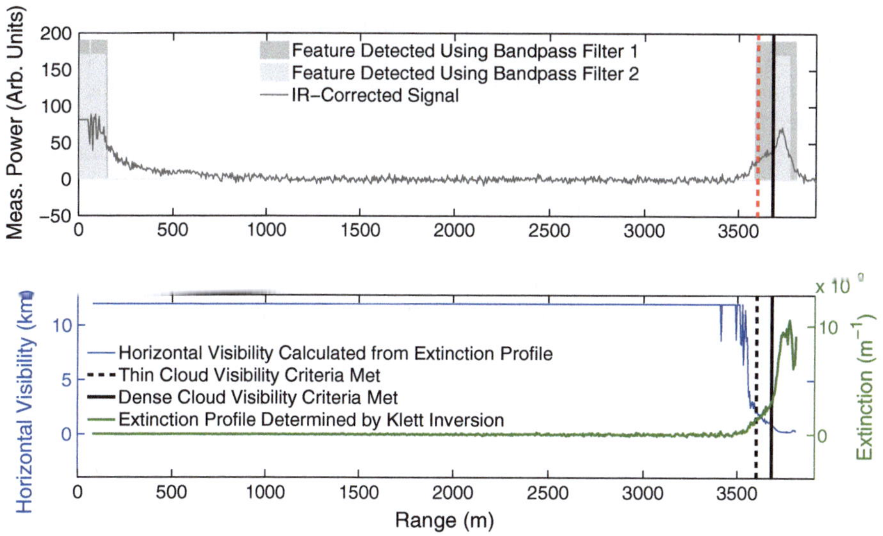

Fig. 5.20 Cloud detection algorithm applied to prototype measurement made 5 June 2012 at 03:16 UTC at Chilbolton Observatory. A vertical *red line* in the *upper graph* indicates the height of cloud base reported by the algorithm

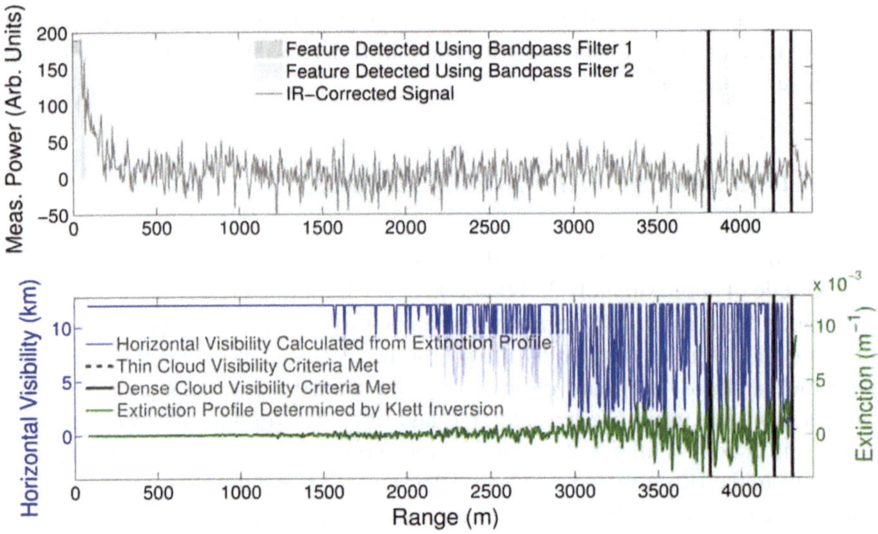

Fig. 5.21 Cloud detection algorithm applied to prototype measurement made 5 June 2012 at 04:25 UTC at Chilbolton Observatory. No cloud was reported by the algorithm

graph reveal the detections to be insignificant. However, returns from just above 3,500 m are determined to be cloud by both methods. The cloud feature as a whole ranging from 3,600 to 2,800 m is detected using bandpass filter 1, while its narrower peak ranging from 3,680 to 3,770 m is detected using bandpass filter 2. Narrow and wide cloud bases found using the visibility method are located at 3,600 and 3,680 m, respectively, as shown in the lower graph and replicated on the upper graph. Since the features run together, one cloud base is output at 3,600 m, indicated by the vertical red dashed line in the upper graph.

Figure 5.21 shows an example from later the same morning when a cloud layer was not reported. In this case instability of the inversion at far ranges due to noise led to the detection of three dense cloud layers using the visibility method as shown in the lower graph. The feature detection method again detected close range noise and/or aerosol returns as shown in the upper graph. The lack of any co-located cloud detections identified by both methods meant that no cloud layer was identified based on this measurement.

5.7.2 Proximity Margins and Rounding Rules

Occasionally a cloud layer may meet the criteria for both of the methods, but the feature detection indicated by the bandpass filtering method may be slightly offset in range from the point indicated by the visibility threshold method. In order to allow for such offsets and avoid the non-reporting of detected cloud layers, small proximity margins have been introduced. Below 1.5 km a proximity margin of 30 m is allowed between the detections flagged up by the two methods; the lower of the two detections is reported as the cloud base. Above 1.5 km a proximity margin of 60 m is applied in a similar manner, and again the lower of the two detections is taken as the cloud base. Note that a thin cloud identified using one method located within the margin of a dense cloud identified using the other method is still considered to be a cloud.

According to the World Meteorological Organization, cloud layers are to be reported such that they may be easily implemented into standardised codes called METAR codes as described in the *WMO Manual on Codes* [29]. This document states that for aviation applications, cloud base heights must be reported at the very least to up to 1,500 m, the top of the operationally-critical zone. Recall from Chap. 1 that most instruments report to significantly greater heights than this, and from Chap. 3 that the FAA in the US requires bases to be reported at least up to 3,810 m. The WMO states that from 30 to 3,000 m, heights should be reported in multiples of 30 m and rounded down to the nearest 30 m range bin.

5.7.3 Cloud Detection Output Comparison

Cloud base height reports from the prototype instrument over a 24 h period at Chilbolton Observatory on 5 June, 2012 are shown in Fig. 5.22. Throughout the first half of the day a gradually descending cloud layer was observed, with a second, lower layer appearing after 07:00. After around 12:00, a very low, stable cloud layer was reported throughout the rest of the observation period.

This was compared with cloud base reported on the same day by the Chilbolton CT75k research ceilometer, the outputs of which are shown in Fig. 5.23. Cloud layers were tracked similarly by both instruments throughout the day, though the CT75k was slightly more sensitive to upper layers in the presence of lower layers, for example, from 12:00 to 13:00 UTC, probably due to a more sensitive tuning of its algorithm. Comparison of the number of lowest layer cloud hits by the two instruments revealed that the CT75k reported at least one layer 2,578 of the 2,880 measurement cycles taking place every 30 s throughout the 24 h comparison period, while the prototype

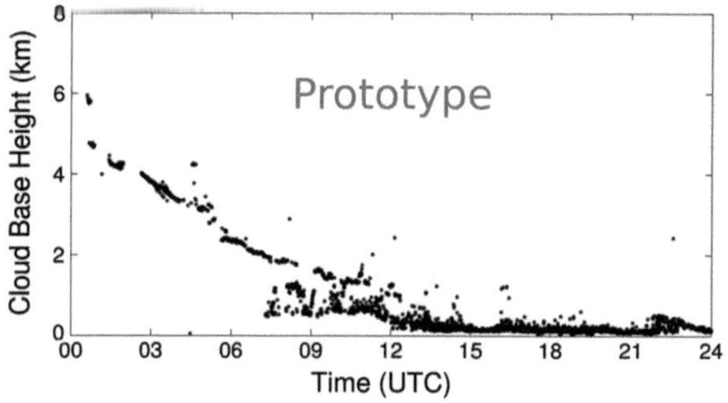

Fig. 5.22 Cloud base determined by the prototype algorithm run on data gathered by the prototype instrument at Chilbolton Observatory on 5 June 2012

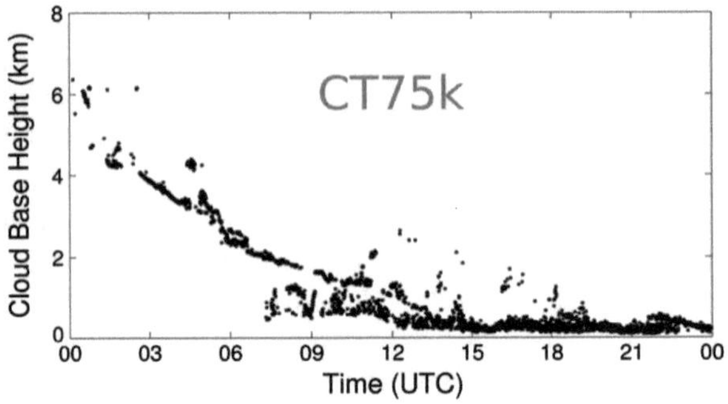

Fig. 5.23 Cloud base determined by Vaisala CT75k research ceilometer at Chilbolton Observatory on 5 June 2012

instrument reported at least one layer 2,232 times. This equates to cloud detected 89.5 % of the time throughout the day by the CT75k and 77.5 % of the time by the prototype. The prototype algorithm could be tuned for more sensitivity by decreasing the thresholds for bandpass filtered feature detection, but this might mean that insignificant cloud layers would be reported, which is undesirable for aviation applications.

To compare the two instruments in terms of their relative cloud base height reports, the lowest layer output throughout the day was compared in cases where the two instruments deviated by less than 300 m in order to ensure the instruments were reporting the same layer, with the prototype outputs rounded down to the nearest 10 m mark to match the 10 m reporting resolution of the CT75k. The two instruments gave positive 1st layer reports within 300 m of each other 1,931 times out of the 2,880 measurement cycles during the 24 h period. On average the prototype instrument reported bases 80 m lower than the CT75k, with the standard deviation of this difference being 78 m and the mode being 40 m. A linear regression of the same filtered data, that is, lowest cloud bases reported within 300 m of each other simultaneously by both instruments, is shown in Fig. 5.24, where the least squares linear fit has an R squared value of 0.998. The y-intercept of linear fit is located at 91 m, and the

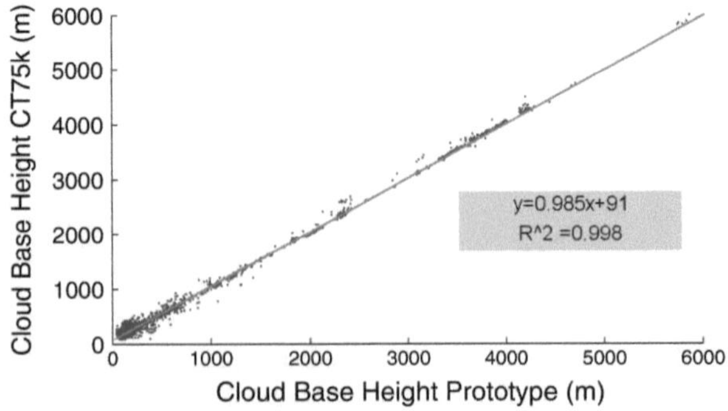

Fig. 5.24 Linear regression of lowest cloud base layer height detections from Figs. 5.22 and 5.23. Layers were required to be within 300 m of each other for this analysis in order to remove comparisons of reports from different layers

Fig. 5.25 Hourly sky images from camera-axis 2100 at Chilbolton Observatory on 5 June 2012 from 03:00 UTC to 14:00 UTC

slope is 0.985, which means the difference between the cloud base reports of the two instruments gradually decreases with height, with the offset becoming essentially zero at a range of 6,000 m.

The CT75k was used as a reference for comparison with the prototype instrument, not as a standard. Recall from earlier in this chapter that Eberhard [7] found that pilots at a 3 degree approach angle typically reported cloud base 79 m lower than the peak of lidar signals for non-precipitating clouds ranging from 160 to 3,200 m. The results of this comparison suggest that it may be possible that the CT75k is reporting cloud base based on peak detection and that the prototype goal of the instrument of assigning cloud base based on visibility aligns with pilot observations in Eberhard's study. Further knowledge of the CT75k algorithm as well as additional intercomparison between these two instruments, particularly at higher ranges, would be necessary to validate such a claim.

In order to provide a sense of what the clouds looked like over the course of the day of these measurements, a series of photos taken at the observatory are shown in Fig. 5.25. Here hourly photos are included from first daylight until 14:00 UTC, after which time the appearance of the sky changed very little for the rest of the daylight hours. At 04:00 it appears that an altocumulus layer at just above 4,000 m is showing through a gap in the gradually descending altrostratus layer which is being reported at 3,500 m by the ceilometers at that time. At 10:00 and 11:00 the altocumulus layer is no longer visible, but the previous altostratus layer has descended to around 1,750 m and would now be categorised as stratus. At the same time a scattered stratocumulus layer is visible below. Occasional gaps in these layers, for example at 12:00, allow the ceilometers to reveal the presence of a third layer, but the photos show primarily the stratocumulus layer at around 200 m with some view of the stratus layer above.

5.8 Determination of Vertical Visibility

If no clouds are detected by a ceilometer, it is expected to report vertical visibility. A recently-introduced international standard, ISO 28902-1 [25], describes an algorithm for lidar determination of visibility for meteorological optical ranges (MOR) of up to 2,000 m, where MOR is defined as the range at which the integrated extinction reaches 3 (a contrast of 5 %). Since the prototype instrument is intended for commercialisation, this standard was followed closely in the development of the instrument's visibility algorithm. The steps of the method described in the standard are summarised in Table 5.1.

An initial far boundary value of $\alpha(r_f) = 1 \times 10^{-1}$/m, corresponding to a visual range of 30 m, is assigned by the standard as the maximum resolvable extinction coefficient for a typical lidar. This value of $\alpha(r_f)$ is lowered through iteration until it is within 10 % of the average extinction α_{ave} calculated from the inversion. A minimum resolvable extinction coefficient $\alpha_{\text{min}} = 1.5 \times 10^{-3}$/m, corresponding to a visual range of 2,000 m, is specified by the standard, and during each iteration any extinction values $\alpha(r)$ below this minimum are omitted from the average α_{ave} that is used to assign the next boundary value $\alpha(r_f)$. This prevents gaps in the scattering media from artificially reducing the boundary value of $\alpha(r_f)$. Once the iteration criteria have been met and the inversion has been completed, the MOR is determined by integrating

Table 5.1 Outline of the ISO lidar vertical visibility algorithm [25]

1. Find r_f, the farthest range at which SNR $\geq$ 6 dB
2. Set initial boundary value of the extinction $\alpha(r_f) = 1 \times 10^{-1}$/m
3. Perform a Klett inversion to determine extinction $\alpha(r)$ (from range r_0, where overlap $O(r) = 0.8$, to range r_f)
4. Determine the system's minimum resolvable extinction, α_{min} (for the prototype $\alpha_{\text{min}} = 1.5 \times 10^{-3}$/m)
5. Find the average α_{ave} of all $\alpha(r) \geq = \alpha_{\text{min}}$
6. If $\lvert\frac{\alpha_{\text{ave}}-\alpha(x_f)}{\alpha_{\text{ave}}}\rvert \geq 0.1$, set $\alpha(r_f) = \alpha_{\text{ave}}$
7. Iterate inversion until $\lvert\frac{\alpha_{\text{ave}}-\alpha(x_f)}{\alpha_{\text{ave}}}\rvert < 0.1$ or an iteration limit is reached
8. MOR is determined from $\alpha(r)$ by $\int_{r_0}^{\text{MOR}} \alpha(r)dr = 3$

$\alpha(r)$ up to the range at which the integral equals 3, and this range is reported as the vertical visibility by the instrument. This iterative method for determination of the boundary extinction value $\alpha(r_f)$ operates under the assumption that when visibilities are low enough to be reported, the extinction coefficient value at the boundary of the region will be similar to the values within the region, i.e., the scatterers in the region and their concentrations are expected to resemble each other, except within clear gaps. Note that according to the standard the accuracy of visibility measurements made including ranges prior to 80 % overlap is questionable due to uncertainties arising from multiple scattering.

Although the most desirable conditions for initial testing of a vertical visibility algorithm would be a uniform fog or haze, these conditions were not encountered during the preliminary test phase, so a precipitation event was selected as a test case. Measurements were taken during a sudden burst of rain that lasted for about 1 minute. Images from the observatory taken just before and during the rain are shown in Fig. 5.26. A droplet counting rain gauge (*droplet counter b*) located at the observatory reported a rainfall of 0.56 mm during this time. A ground-level visibility of 2,150 m (corresponding to an extinction of $\alpha = 0.0014\,\text{m}^{-1}$) was reported for the same minute by the observatory's Vaisala PWD21 present weather sensor (*cfarr-pwd21*). The CT75k reported vertical visibilities of 210 and 180 m at around the same time.

Figure 5.27 shows the result of the iterative inversion method applied to this low cloud and rain situation along with some parameters calculated from it. The integrated extinction was 2.38, indicating a contrast of 9.3 %. Therefore the integrated extinction threshold of 3 corresponding to a contrast ratio of 5 % dictated by the standard [25] was not reached. Since that threshold was not reached, the average extinction over the measurement region of significant returns was used to calculate an average visibility of 196 m, which turned out to be close to the two visibility values of 180 and 210 m reported by the CT75k. While the standard should be followed for the primary visibility output, a secondary output giving average visibility and the depth of the averaging region could prove useful in cases where visibility is nearly obscured but the threshold is not met. The usefulness of this approach was examined as follows.

Fig. 5.26 Images taken by horizontal camera at Chilbolton Observatory at 10:45:15 UTC (*left*) and 10:46:00 UTC (*right*) on 29 August, 2012, just before and during a sudden burst of rain that reduced visibility for approximately 1 min

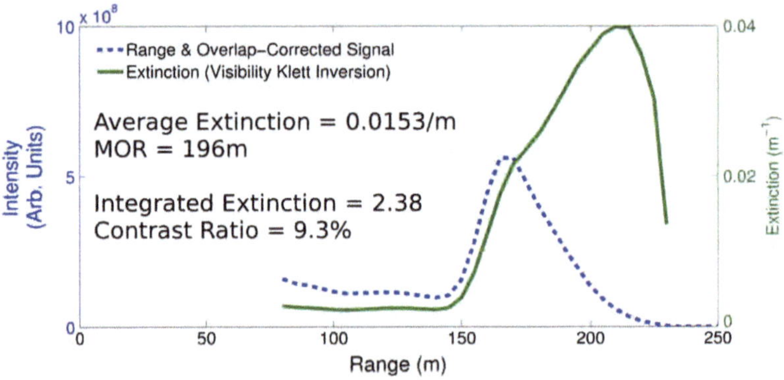

Fig. 5.27 Range-corrected and overlap-corrected prototype signal and its inversion by application of the visibility algorithm iterative method at 10:46 UTC on 29 August, 2012 at Chilbolton Observatory

While averaging the extinction coefficient over the region from 80 m (where overlap begins to be significant) to the top of the signal as shown in Fig. 5.27 and using this to calculate the visibility gave results that appear to be in line with the CT75k visibilities, this does not accurately reflect the visibility situation during this measurement. This is because in this case, two features are contributing to obscuration of visibility: the rain and the low cloud. The cloud, the base of which is located at a height of 150 m, provides significantly more extinction than the rain. Therefore, averaging only over the region from 80 m and above biases the extinction to a higher value, and therefore the visibility to a lower value of 196 m. Since the extinction appears to be fairly uniform in the rain, in these special circumstances the range and overlap-corrected signal was extrapolated in order to extend it to ground as shown in Fig. 5.28.

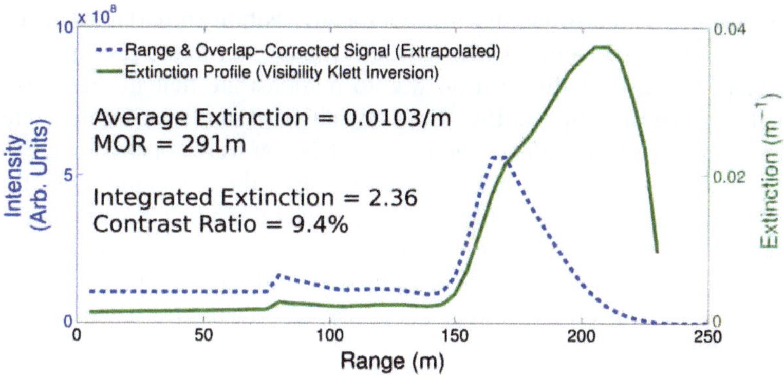

Fig. 5.28 Extrapolated range-corrected and overlap-corrected prototype signal and its inversion by application of the visibility algorithm iterative method at 10:46 UTC on 29 August, 2012 at Chilbolton Observatory

Fig. 5.29 Image taken by vertical camera at Chilbolton Observatory at 10:30:01 UTC on 29 August, 2012

In this case, the average extinction coefficient that is calculated from the signal is somewhat lower, and therefore gives a higher visibility of 291 m. The photograph in Fig. 5.29 indicates that the cloud layer was not fully obscuring and suggests that an observer on the ground could therefore have seen, through the cloud, a target positioned 291 m above the ground. While it may or may not be useful to provide an alternative report of visibility over the range of significant returns, investigation of this method reveals that the extrapolated inverted extinction coefficient near the surface matches closely that reported by the ground-based present weather sensor at the same time, thus providing some validation of the iterative inversion method. This investigation also demonstrates that in order for any visibility report to be meaningful, the definitions and algorithms applied must be made clear to the user of the data.

The main purpose of this section was to demonstrate that the iterative inversion method suggested by the ISO visibility standard could be implemented in a satisfactory manner. While ideal conditions such as fog were not encountered during the preliminary testing phase of the prototype algorithms, suitable conditions for testing the iterative inversion method were encountered. In fact, the presence of rain beneath the cloud allowed the inverted data to be extrapolated to ground level and checked with ground-level readings of extinction. Although the prototype visibility algorithm can be refined further, it has been shown here to meet basic expectations of functionality.

5.9 Conclusion

A method for correcting for phase distortion by the impulse response of the prototype amplifier has been explained. The calibration of attenuated backscatter returns based on prototype signals corrected for impulse response has been performed. After this calibration, attenuated backscatter returns were shown to agree with those from the CT75k research ceilometer at Chilbolton Observatory within a maximum ratio of 1.5 for a variety of different scattering media at a variety of different ranges. On the calibration data itself the instruments agreed within 15 %. In addition, sensitivity of the prototype to boundary layer aerosols has been demonstrated.

A variety of cloud detection algorithms from the literature have been reviewed, and a visibility-based cloud detection algorithm built upon the method of Poyer and Lewis [20] has been developed to determine cloud base height using inverted data. To improve the performance of this method, it has been coupled with a new algorithm based on two bandpass filters that identify cloud signatures in lidar returns. This second method looks for cloud-return-shaped features in the signal after it has been corrected for impulse response, but before it has been range-corrected and inverted. Only when a threshold in the bandpass filtered data is exceeded and the visibility threshold criteria are met in the inverted data at the same range is a cloud reported. These two approaches in the parallel cloud detection algorithm used in the prototype balance each other to remove spurious detections while still providing good sensitivity.

Compared to the CT75k, the prototype has been shown to track layers similarly. The sensitivity of the cloud detection algorithm described in this chapter and applied to the prototype instrument is slightly lower than that of the CT75k. Since the CT75k was used as a reference for cloud base but was not considered to be a standard, this could mean that the prototype algorithm should be tuned for increased sensitivity or that it is simply better at filtering out clouds that are insignificant for aviation applications. For same-layer lowest cloud base height reports (lowest layers where the two instruments reported within 300 m of each other), the prototype algorithm gave outputs on average 80 m below those of the CT75k.

In the case of no cloud report, a vertical visibility output is given by the prototype based upon the method [25] recommended by the International Organization for Standardization. This method was described here and the results of its application to a rain event validated the method's usefulness.

The prototype instrument is therefore capable of providing the three primary outputs required of a ceilometer: cloud base height, vertical visibility, and attenuated backscatter, and has shown sensitivity to boundary layer aerosols.

References

1. S. Jones, The impulse response of amplifiers with six identical single-tuned stages. IEEE Trans. Circuit Theory **10**(2), 302 (1963)
2. E.W. Weisstein, Convolution theorem. From MathWorld-A Wolfram Web Resource, 2012, http://mathworld.wolfram.com/ConvolutionTheorem.html
3. E. O'Connor, A. Illingworth, R. Hogan, A technique for autocalibration of cloud lidar. J. Atmos. Oceanic Tech. **21**, 777–786 (2004)
4. E. O'Connor, Lidar calibration. University of Reading Department of Meteorology. [Online], 2012, http://www.met.reading.ac.uk/~swr99ejo/lidar_calibration/
5. G. Tsaknakis, A. Papayannis, P. Kokkalis, V. Amiridis, H.D. Kambezidis, R.E. Mamouri, G. Georgoussis, G. Avdikos, Inter-comparison of lidar and ceilometer retrievals for aerosol and planetary boundary layer profiling over Athens, Greece. Atmos. Meas. Tech. **4**, 1261–1273 (2011)
6. C.M. Platt, S.A. Young, A.I. Carswell, S.R. Pal, M.P. McCormick, D.M. Winker, M. Del-Guasta, L. Stefanutti, W.L. Eberhard, M. Hardesty, P.H. Flamant, R. Valentin, B. Forgan, G.G. Gimmestad, H. Jger, S.S. Khmelevtsov, I. Kolev, B. Kaprieolev, D.-R. Lu, K. Sassen, V.S. Shamanaev, O. Uchino, Y. Mizuno, U. Wandinger, C. Weitkamp, A. Ansmann, C. Wooldridge, The experimental cloud lidar pilot study (ECLIPS) for cloud-radiation research. Bull. Amer. Meteor. Soc. **75**(9), 1635–1654 (1994)
7. W.L. Eberhard, Cloud signals from lidar and rotating beam ceilometer compared with pilot ceiling. J. Atmos. Oceanic Technol. **3**(3), 499–512 (1986)
8. W.E. Eggert, Approach visibility studies at newark, final report AMB Project D-1-902. Technical Repore, Bureau of Research and Federal Aviation Agency, Washington, D.C., 1960
9. S.R. Pal, W. Steinbrecht, A.I. Carswell, Automated method for lidar determination of cloud-base height and vertical extent. Appl. Opt. **31**(10), 1488–1494 (1992)
10. E. Clothiaux, G. Mace, T. Ackerman, T. Kane, An automated algorithm for detection of hydrometeor returns in micropulse lidar data. J. Atmos. Oceanic Tech. **15**(8), 1035–1042 (1998)
11. J. Spinhirne, Micro pulse lidar. IEEE T. Geosci. Remote **31**(1), 48–55 (1993)
12. D.M. Winker, M.A. Vaughan, Vertical distribution of clouds over Hampton, Virginia observed by lidar under the ECLIPS and FIRE ETO programs. Atmos. Res. **8095**, 93 (1994)
13. C. Platt, D. Winker, M. Vaughan, S. Miller, Backscatter-to-extinction ratios in the top layers of tropical mesoscale convective systems and in isolated cirrus from LITE observations. J. Appl. Meteorol. **38**(9), 1330–1345 (1999)
14. M. Vaughan, S. Young, D. Winker, K. Powell, A. Omar, Z. Liu, Y. Hu, C. Hostetler, Fully automated analysis of space-based lidar data: an overview of the CALIPSO retrieval algorithms and data products. in *Laser Radar Techniques for Atmospheric Sensing, Proceedings of SPIE*, vol. 5575, 2004
15. J.R. Campbell, D.L. Hlavka, J.D. Spinhirne, D.D. Turner, C.J. Flynn, *Operational Processing and Cloud Boundary Detection from Micro Pulse Lidar Data* (ARM Science Team, Tucson, 1998), pp. 119–122

16. J.L. Gaumet, J.C. Heinrich, M. Cluzeau, P. Pierrard, J. Prieur, Cloud-base height measurements with a single-pulse erbium-glass laser ceilometer. J. Atmos. Oceanic Technol. **15**(1), 37–45 (1998)
17. Y. Morille, M. Haeffelin, P. Drobinski, J. Pelon, Strat: An automated algorithm to retrieve the vertical structure of the atmosphere from single-channel lidar data. J. Atmos. Oceanic Technol. **24**(5), 761–775 (2007)
18. Z. Wang, K. Sassen, Cloud type and macrophysical property retrieval using multiple remote sensors. J. Appl. Meteor. **40**(10), 1665–1682 (2001)
19. C. Bohren, *Clouds in a Glass of Beer: Simple Experiments in Atmospheric Physics* (Dover Science Books, Dover Publications, Mineola, 2001)
20. A.J. Poyer, R. Lewis, Adapting the micropulse lidar for use as a reference for cloud measurement. *AMS Fourth Symposium on Lidar Atmospheric Applications*, 2009
21. H. Koschmieder, Theorie der horizontalen sichtweite. Beitr age zur physik der freien Atmosph are., vol. 12, 1924, pp. 33–53
22. W. Middleton, *Vision Through the Atmosphere* (University of Toronto Press, Toronto, 1963)
23. World Meteorological Organisation, International meteorological vocabulary. Technical Report WMO No. 182, 1992
24. International Civil Aviation Organization, *Annex 3: Meteorological Services for International Air Navigation*, 16th edn. (ICAO, Montreal, 2007)
25. I. O. for Standardization (ISO), ISO 28902–1:2012, air quality-environmental meteorology-ground-based remote sensing of visual range, ISO Std
26. J.D. Klett, Stable analytical inversion solution for processing lidar returns. Appl. Opt. **20**(2), 211–220 (1981)
27. W. Carnuth, R. Reiter, Cloud extinction profile measurements by lidar using klett's inversion method. Appl. Opt. **25**(17), 2899–2907 (1986)
28. I. Binietoglou, A. Amodeo, G. D'Amico, A. Giunta, F. Madonna, L. Mona, G. Pappalardo, Examination of possible synergy between lidar and ceilometer for the monitoring of atmospheric aerosols. in *Proceedings of SPIE*, vol. 8182, 2011, p. 818209
29. World Meteorological Organisation, WMO manual on codes. Technical Report Publication No. 306, 2011

Chapter 6
Conclusions and Further Work

6.1 Conclusions

A novel lidar ceilometer prototype has been designed, built, and tested. The design has an optical signal to noise advantage of greater than 4 times that of a leading instrument in a similar class. A divided-lens design with close proximity of optical axes efficiently utilises the elliptical output of a diode laser without the need for beam-correcting optics, balances relatively fast turn-on of the overlap function (overlap is greater than 80 % at a range of 200 m) with good optical isolation between the channels, and, since two halves of the same lens are used, the transmitter and receiver can be focused in a single procedure. A compact mechanical design for the optical assembly has been produced, and a straightforward optical alignment procedure has been devised. Overall, low cost components have been selected, the number of components has been minimised, and the assembly process simplified as much as possible. The goal of this design of improving the optical performance of ceilometers while at the same time reducing their cost has been achieved through an effective compromise between these two competing approaches, increasing the range of the instrument by approximately 25 % while decreasing the estimated selling price by approximately 25 % compared to commercially available instruments in a similar class.

As a good understanding of the close range sensitivity of this type of instrument is important for cloud detection and particularly important for aerosol profiling and determination of vertical visibility, a novel experimental method for the characterisation of the range-dependent transmitter-receiver overlap has been described and applied to the prototype. This method utilises an imaging technique which presents a virtual image of a scattering target to a lidar instrument at various effective ranges in order to characterise the overlap of the system throughout its entire range within the limited space of a laboratory. In addition, a geometric calculation of the overlap has been derived specifically for the unique geometry of the prototype system. This calculation has a great deal of built-in flexibility for dealing with different laser emitter patterns and aperture configurations and is therefore a useful tool for lidar design intercomparison. Finally, to provide another reference, the overlap of the

J.D. Vande Hey, *A Novel Lidar Ceilometer*, Springer Theses,
DOI 10.1007/978-3-319-12613-5_6

prototype was characterised by measuring returns from a hard target translated along the overlap region.

The hard target measurements fell within the error bars of the imaging-based overlap measurement beyond the 50 % overlap range of approximately 125 m, but not below it. The reasons for the differences between these two measurements at close ranges were not understood quantitatively, but a number of potential causes were identified, including possible inadvertent shift in the optical alignment of the prototype between the two test sites. The calculated overlap fell within the experimental error bars on the imaging based overlap measurement in the region from 125 to 200 m corresponding to approximately 50–80 % overlap. As anticipated, however, calculation and experiment did not agree entirely. In the overlap onset region and in the region approaching full overlap there were nontrivial differences. Poorly understood optical effects not accounted for in the calculation, such as inadvertent scattering and/or unintentional laser tilt, were the most likely causes of these discrepancies. The differences between measured and calculated overlap underscore the importance of measuring overlap prior to deploying a lidar instrument in order to evaluate the performance of its optical design.

Once the prototype had been built and its overlap had been characterised, it required calibration as well as the development of automated signal processing methods in order for the instrument to be capable of reporting the necessary outputs of attenuated backscatter, vertical visibility, and cloud height.

The instrument was given a preliminary calibration following a method described in the literature. Attenuated backscatter calibration measurements were found to agree to within 15 % of those of a research ceilometer calibrated by the same method. In addition the instrument has demonstrated sensitivity to boundary layer aerosols and the attenuated backscatter profile can clearly be used to monitor the vertical extent of concentrated aerosol layers in this region for lidar-based boundary layer height retrieval.

A method to allow the prototype to automatically determine vertical visibility from the signal has been implemented. This method, based on an iterative inversion algorithm described in the literature, has been tested and shown to provide reasonable results through comparison with outputs from a research ceilometer and visibility measured by a ground-based present weather sensor.

A novel two-part algorithm has been developed for the automated detection of clouds. This algorithm uses two sets of criteria in order to identify cloud base. The first set of criteria is based on a method described in the literature which assigns cloud base where certain threshold conditions, expressed separately for dense and diffuse clouds in terms of horizontal visibilities derived from the inverted extinction profile, are satisfied. The second set of criteria is based on feature identification in the non-range-corrected signal. In this second method, potential cloud return features are located by application of two bandpass filters which are designed to filter out high frequency noise and very broad low frequency features such as precipitation returns. Only when a feature meets both sets of detection criteria is a cloud layer reported by the prototype algorithm. Since the visibility-threshold method is somewhat oversensitive at far ranges where the inversion may be unstable, and the bandpass filter method is somewhat over-sensitive at near ranges due to noise artefacts and increased sensitivity to aerosol returns, the effect of using these two methods together is that false reports are to a large extent avoided. Cloud outputs from the prototype instrument tracked closely those from a research ceilometer used as a reference, showing slightly less sensitivity (77.5 % compared to 89.5 % positive detections of at least one layer during the measurement period) and slightly lower reporting (on average 80 m lower) than the research instrument. These differences in the prototype's sensitivity and cloud base height reports could be advantageous for aviation, but further investigation is required to validate this supposition. Regardless of the differences between the two instruments, the prototype algorithm was shown to be fully functional for tracking cloud base.

By lowering the cost of lidar ceilometers, improving their performance, and deploying the technology broadly, more coverage of the Earth can be made, and more people can contribute to the meaningful study of the atmosphere through widespread networks. Not only large scale government agencies such as the Met Office, but also schools, local governments, universities, and amateur weather associations can become involved in this process, a process that can help to create a deeper understanding of climate and weather on many levels.

6.2 Further Work

There are many exciting and challenging areas of research that can be explored in order to build upon the current work. The primary recommendations for further study fall into three categories: Overlap calibration and correction, signal processing methods, and instrumentation.

6.2.1 Potential Advances in Overlap Characterisation and Correction

6.2.1.1 Portable Overlap Characterisation Tool

There are three particularly important areas to investigate following the work that has been described in this thesis regarding measurement of overlap. The first of these is the implementation of the laboratory-based compound imaging system described in Chap. 4 into a portable system for the calibration of elastic lidar overlap. Though it would require significant effort in terms of optical design and alignment, it could, for example, be used to characterise the overlap of ceilometers coming off an assembly line or used at observatories or in the field to calibrate elastic systems. The main obstacle to the development of a compact, mobile instrument would be the cost of imaging optics with large aperture and suitably low $f/\#$. This requirement is exactly the same as that of the imaging system in a compact, large aperture lidar, however, and although expensive, should not be prohibitive.

6.2.1.2 Multiple Scattering Models for Maximising Usefulness of Signal Prior to Full Overlap

A second important problem relating to overlap is the question of how to gain useful information from the signal prior to full overlap. Based on good understanding of the overlap, and if some assumptions about the particles in the closest part of the atmosphere can be made, a multiple scattering model could be used for inverting close-range returns. For the prototype instrument, signals due to multiply-scattered returns have been detected as close as 10 m to the instrument. With another instrument as a reference to provide particle size information, or possibly with the help of a model, a multiple scattering calculation method such as that of Eloranta [1] could be used to estimate concentrations of particles prior to full overlap. For the prototype instrument, this could improve visibility measurements, low cloud detection, and aerosol profiling. While the errors that would likely result through the use of such a method might be too high for detailed aerosol study with research lidars, this approach could be extremely useful for improving the quality of close-range ceilometer data.

6.2.1.3 Multiple Fields of View

The third area worth investigating is application of multiple fields of view, for example, by implementing a method described by Hutt et al. [2] which utilised a detector based on isolated concentric detection regions. This could provide a close range wide field of view channel with fast overlap and a far-range narrow field of view channel with slow overlap. This approach might also facilitate determination of the lidar ratio and effective particle radii if, as noted by Hutt et al., signals were detected simultaneously by three different field of view channels. Implementation of this method would require significant consideration of detector design, but would likely at the very least provide a good solution to the near-range overlap versus far-range SNR tradeoff.

6.2.2 Signal Processing Approaches for Full Exploitation of Ceilometer Potential

Further fine-tuning of the prototype algorithm through implementation of techniques such as variable cloud detection thresholds based on time series analysis of feature persistence is certainly possible. But beyond this kind of work there are a number of signal processing methods that should be studied in an attempt to increase the value of the prototype's data.

6.2.2.1 Additional Outputs

First, there are two additional outputs that should be considered. According to Chiu et al. [3], optical depth is the most important property of clouds in terms of their influence on the Earth's radiation budget, but is unfortunately also one of the most poorly observed. While the upper limit of optical depth determined by a ceilometer would be not much greater than $\tau = 3$, with a good inversion, optical depths of thin clouds could be tracked by ceilometers, and clouds that fully extinguish the signal could simply be reported as being above a threshold. Since ceilometers are widely deployed, this could contribute to the effort of characterising cloud optical depth around the globe.

Another output that should be considered is height-dependent cloud fraction. Ceilometers are often expected to report an estimate of cloud cover in terms of oktas or eighths of the sky in order to match what a visual observer might report. Bretherton et al. proposed an alternative approach that is a more appropriate for a vertically-pointing instrument. They applied a method whereby cloud bases were measured each minute, and on an hourly basis cloud fraction was calculated based on these measurements in each of a number of height ranges [4]. This is a much more useful way to evaluate ceilometer data because it utilises the vertically-resolved capability of the measurement and doesn't require horizontal extrapolation that varies widely in quality depending upon cloud structure and wind velocity. It is strongly recommended that this approach be used over the traditional sky condition report, and it is hoped that meteorological agencies will soon adopt this approach.

6.2.2.2 Improving Boundary Layer Aerosol Profiling

As discussed in Chap. 1, monitoring of planetary boundary layer aerosols is an important application of ceilometers. Due to the somewhat noisy signals of ceilometers and the somewhat difficult assumptions that are often necessary for a good inversion of the signal, aerosol profiling by these instruments is typically limited to boundary layer height retrieval. The variable spatial and temporal averaging approach of Stachlewska et al. [5] could prove to be be a useful tool for improving the quality of aerosol retrievals from ceilometers. Complexities in the boundary layer, however, can mean that the boundary layer top reported by a gradient method might not correspond to the structure of the temperature inversion. The method of Di Giuseppe et al. [6], in which measured data are evaluated using a boundary layer model, appears to be the way forward in terms of getting the best possible information from ceilometer measurements. The prototype that has been developed should be used to investigate the performance of these two methods used together. The powerful processor in the prototype electronics could be used to do some of the required processing inside the instrument on a quasi-real time basis, particularly if model data were remotely accessible.

In addition, the quality of ceilometer aerosol data can be further improved through synergy with complementary, co-located instruments, such as sun photometers or radiosondes. Research in this area has already begun, as evidenced by the work of Madonna et al. [7], Heese et al. [8], and Tsaknakis et al. [9]. Also, networks of ceilometers can provide good coverage of a geographical area to study aerosol transport in the boundary layer as described, for example, by Englebart et al. [10]. To further extend the usefulness of the ceilometer described here, it should be installed in such networks, and its attenuated backscatter data output should be fully characterised and calibrated to ensure accuracy. In addition, efforts should be made to establish the data gathering infrastructure required to record and evaluate attenuated backscatter profiles from ceilometers based at airports and helipads around the world as currently most of these simply report cloud base height and vertical visibility in METAR format.

6.2.2.3 Forward Modelling of Ceilometer Signals

Another area that should be considered for the evaluation of ceilometer signals is forward modelling. In a forward modelling approach, rather than inverting the signal, which for ceilometers can vary significantly in accuracy depending upon available knowledge of the local atmosphere, an attenuated backscatter signal is generated from an atmospheric model and compared with the signal itself, and then data assimilation can be used to further improve the accuracy of the modelled conditions. Since the quality of modelled data is constantly improving, this can be an extremely effective way to utilise ceilometer data. This is a current area of interest at the UK Met Office, where preliminary work has been done by Cox and Charlton-Pérez [11]. In order for this to be implemented in a useful manner, a network of well-characterised, well-calibrated ceilometers is necessary. In order to pursue research in this direction, a network of ceilometers based on the prototype should be installed and their outputs assimilated into a model.

6.2.3 Further Development of Instrumentation

6.2.3.1 Diffraction-Limited Optics

While the prototype instrument has met its design specification at a reasonable cost, there are a number of follow-on instruments that could be developed in order to further extend the capabilities of this instrument. The first area to consider is the incorporation of diffraction-limited optics and a single-mode laser source. The diffraction limited spot radius for an optical system with an aperture radius of 75 mm and an $F/\# = 2$ is 2.5 μm. This would allow a much smaller receiver field of of view of 7.5 μrad and therefore improve the daytime SNR by a factor of 100 times for the same eye-safe laser power. In addition, the wavelength bandwidth of the single-mode laser would be significantly narrower, which would allow a narrower bandpass filter to be used. If a filter with a bandwidth of 0.2 nm were used (the bandwidth of the

Micropulse Lidar [12]), SNR would further increase by a factor of 13.4. Despite the costs and complexity involved, and the fact that full overlap height would be significantly higher, the move to a high quality laser and diffraction-limited optics would increase SNR by 3 orders of magnitude over the current system. The challenge here is to find the best, most current, cost effective technology, and implement it in an effective manner in order to achieve an instrument that could perform similarly to the Micropulse Lidar but at a significantly lower cost.

6.2.3.2 Depolarisation

One modification to the instrument that would be valuable to both the research and meteorology communities would be the introduction of a depolarisation channel. This would allow a user to distinguish ice clouds from water clouds, but also to identify the sphericity of aerosols. Depolarisation for identification of volcanic ash clouds is particularly important for aviation. This capability has been included on the Micropulse Lidar [12], but has not yet been incorporated into a low cost ceilometer. Depolarisation could be achieved on the prototype by using a polarising beamsplitter and two detector channels.

6.2.3.3 Multiple Laser Wavelengths

Another possible extension of the prototype instrument would be the incorporation of a second wavelength, for example at 1,550 nm, to allow particle radii to be determined by analysis of the colour ratio between two wavelengths. This could be achieved using a dichroic beamsplitter/combiner.

6.2.3.4 Doppler Profiling

Doppler lidar for wind speed is another technology that can be implemented into manufactured lidar systems. According to Werner [13], the preferred method for Doppler lidar measurements is the heterodyne mixing technique which resolves the difference between the transmitted signal frequency and the return signal frequency. This however, requires an extremely stable laser source. Commercial Doppler lidar systems capable of measuring wind speed are already available, for example, from Leosphere [14], so in order to justify work in this area, a significant technological advance or significant cost-savings would be required.

6.2.3.5 Differential Absorption

An extended research area worthy of serious consideration is low-cost differential absorption lidar. Differential absorption lidar (DIAL) probes the atmosphere for concentrations of specific atmospheric molecular constituents by transmitting two wavelengths: one "online" wavelength that is a specific absorption wavelength of the species of interest and one "offline" wavelength that is not absorbed by the species. By analysing the ratio of these two, the concentration of the specific atmospheric constituent can be determined. The DIAL technique has been applied to a wide variety of trace gases including industrial emissions such as O_3, NO_2, SO_2, NH_3, HCl, CO, and Hg and has become a powerful tool for air quality monitoring [15]. DIAL systems require laser wavelengths at the specific absorption lines of the species. This means that obtaining an appropriate laser can be difficult unless a tunable source is available. In addition, there are tight tolerances on the optical filter, and high sensitivity requirements on the detector. As with some of the other technologies discussed here, it is a difficult but important task to reduce the cost of DIAL instruments, particularly because there are so many chemical species of interest that can be studied by this technique. Due to its broad absorption spectrum and the fact that it is highly absorbing of solar background light, O_3 might be the starting point for this kind of work.

6.2.3.6 High Spectral Resolution

Perhaps the most exciting further work in instrumentation, from both scientific and engineering perspectives, would be the development of a low-cost high spectral resolution lidar (HSRL). While Rayleigh-scattered returns from molecules show temperature-dependent Doppler broadening based on a distribution of molecular velocities which varies depending on the kinetic energy of the gas, Mie-scattered returns from larger, slower moving aerosols, exhibit significantly lower temperature-dependent spectral shift. According to Eloranta [16], HSRL instruments typically distinguish aerosol from molecular returns by resolving a narrow spectral linewidth with a Fabry-Perot interferometer, either in a scanning or a fixed configuration. Rayleigh-scattered returns from the atmosphere can be predicted from a model in order to calibrate the spectrum, and the independently-resolved Mie and Rayleigh signals can then be used to determine the lidar ratio at every range of the instrument without the need for additional information.

In order to build a "low-cost" HSRL, however, there are a number of challenges. First, there are tight specifications on some of the components. For example, a highly stable single-mode laser with high spectral purity (linewidth below 100 MHz) is required, along with a matched, stabilised bandpass filter (bandwidth below 1 GHz). Secondly, the complexity of the system can be great. According to Eloranta, the University of Wisconsin's Arctic HSRL uses more than 50 optical components in order to achieve the stable spectral resolution required to distinguish the Doppler-broadened Rayleigh return from the Mie Return. However, with continuing improvements in available laser and optics technology, and with considerable engineering effort, it should become possible to build a compact, high spectral resolution lidar at an affordable cost. Due to its ability to determine the lidar ratio, this type of instrument would see broad deployment in the field.

6.2.3.7 Short-Range Aerosol Profiler

In contrast to the high spectral resolution lidar, which is extremely technology-intensive, there is an area of research in the other direction, straightforward and inexpensive, that is well worth investigating. While research lidar systems, and even ceilometers, typically have substantial ranges, this often means that data quality near the ground is sacrificed for improved signal to noise ratios at far ranges. An instrument that could be of importance to the research and meteorological communities is a short-range, wide field of view lidar. An instrument such as this, with a very short full overlap distance, could be used to profile aerosols near the boundary layer, perhaps up to around 1,000 m. This is something that could be implemented based on the prototype technology, and is recommended as the first step in further work relating to instrumentation.

6.3 Conclusion

All of the these possibilities for further work emerging from the current research are worth pursuing. While some of them have direct commercial implications and others do not, they are all interesting and significant from a scientific perspective. The prototype lidar instrument that has been developed and the techniques that have been employed in its implementation and characterisation provide a substantial base upon which to build further research.

References

1. E.W. Eloranta, Practical model for the calculation of multiply scattered lidar returns. Appl. Opt. **37**(12), 2464–2472 (1998)
2. D.L. Hutt, L.R. Bissonnette, L. Durand, Multiple field of view lidar returns from atmospheric aerosols. Appl. Opt. **33**(12), 2338–2348 (1994)
3. J.C. Chiu, C.-H. Huang, A. Marshak, I. Slutsker, D.M. Giles, B.N. Holben, Y. Knyazikhin, W.J. Wiscombe, Cloud optical depth retrievals from the aerosol robotic network (AERONET) cloud mode observations. J. Geophys. Res. **115**(D14), 202 (2010)
4. C.S. Bretherton, E. Klinker, A.K. Betts, J.A. Coakley, Comparison of ceilometer, satellite, and synoptic measurements of boundary-layer cloudiness and the ECMWF diagnostic cloud parameterization scheme during ASTEX. J. Atmos. Sci. **52**(16), 2736–2751 (1995)
5. I. Stachlewska, M. Piadłowski, S. Migacz, A. Szkop, A. Zielińska, P. Swaczyna, Ceilometer observations of the boundary layer over Warsaw, Poland. Acta Geophys. **60**, 1386–1412 (2012). doi:10.2478/s11600-012-0054-4
6. F. Di Giuseppe, A. Riccio, L. Caporaso, G. Bonaf, G.P. Gobbi, F. Angelini, Automatic detection of atmospheric boundary layer height using ceilometer backscatter data assisted by a boundary layer model. Q. J. R. Meteorol. Soc. **138**(664), 649–663 (2012)
7. F. Madonna, G. Pappalardo, A. Amodeo, G. DAmico, L. Mona, A. Giunta, A. Boselli, V. Cuomo, CNR-IMAA experimental field for atmospheric research, in *Proceedings of the 8th*

International Symposium on Tropospheric Profiling, ed. by W.M.A. Apituley, H.W.J. Russchenberg (2009), pp. S05–P02
8. B. Heese, H. Flentje, D. Althausen, A. Ansmann, S. Frey, Ceilometer lidar comparison: backscatter coefficient retrieval and signal-to-noise ratio determination. Atmos. Meas. Tech. **3**(6), 1763–1770 (2010)
9. G. Tsaknakis, A. Papayannis, P. Kokkalis, V. Amiridis, H.D. Kambezidis, R.E. Mamouri, G. Georgoussis, G. Avdikos, Inter-comparison of lidar and ceilometer retrievals for aerosol and planetary boundary layer profiling over Athens, Greece. Atmos. Meas. Tech. **4**, 1261–1273 (2011)
10. D. Engelbart, J. Reichardt, G. Teschke, Intercomparison of methods for determining mixing height using a new network of single-photon-counting high-sensitivity ceilometers in Germany, in *Proceeding of WMO Meeting, December, 2008*, St. Petersburg, Russian Federation, 2008
11. O. Cox, C. Charlton-Pérez, A met office forward operator for attenuated backscatter, in *Reviewed and Revised Papers of 26th ILRC International Laser Radar Conference*, Porto Heli, Greece, 25–29 July 2012
12. J.R. Campbell, D.L. Hlavka, E.J. Welton, C.J. Flynn, D.D. Turner, J.D. Spinhirne, V.S. Scott, I.H. Hwang, Full-time, eye-safe cloud and aerosol lidar observation at atmospheric radiation measurement program sites: instruments and data processing. J. Atmos. Oceanic Technol. **19**(4), 431–442 (2002)
13. C. Werner, Doppler wind lidar, in *Lidar*, ed. by C. Weitkamp. Springer Series in Optical Sciences, vol. 102 (Springer, Berlin, 2005), pp. 325–354
14. Leosphere, WINDCUBEv2 Brochure, http://leosphere.com/products2.php?cat=wl&rubrique=86. Accessed Sept 2012
15. G. Gimmestad, Differential-absorption lidar for ozone and industrial emissions, in *Lidar*, ed. by C. Weitkamp. Springer Series in Optical Sciences, vol. 102 (Springer, Berlin, 2005), pp. 187–212
16. E. Eloranta, High spectral resolution lidar, in *Lidar*, ed. by C. Weitkamp. Springer Series in Optical Sciences, vol. 102 (Springer, Berlin, 2005), pp. 143–163

Joshua D. Vande Hey Curriculum Vitae

Education

B.A. in Physics, Lawrence University, Appleton, WI, USA, 2001.

Ph.D. in Optical Engineering, Wolfson School of Mechanical and Manufacturing Engineering, Loughborough University, Loughborough, Leics, UK, 2013.

Professional Experience

NERC Knowledge Exchange Fellow in Aerosols and Health
University of Leicester, Leics, UK, 2014-present.

PIKT Associate in Airborne Instrumentation
University of Leicester, Leics, UK, 2013–2014.

Scientific Consultant in Optical Engineering
2011-present.

Optical Engineer
Campbell Scientific, Ltd., Shepshed, Leics, UK, 2009–2010

KTP Associate in Opto-mechanical Engineering
Wolfson School of Mechanical and Manufacturing Engineering, Loughborough University, Loughborough, Leics, UK, in collaboration with Campbell Scientific, Ltd., 2007–2009.

Research and Develpment Engineer in Laser Characterisation Systems
Alfalight, Inc., Madison, WI, USA, 2001–2004.

J.D. Vande Hey, *A Novel Lidar Ceilometer*, Springer Theses,
DOI 10.1007/978-3-319-12613-5

Publications

F. Madonna, F. Amato, J. Vande Hey, G. Pappalardo, "Ceilometer aerosol profiling versus Raman lidar in the frame of INTERACT campaign of ACTRIS," Atmos. Meas. Tech. Discuss, Manuscript Number amt-2014-332, submitted for review, October 2014.

J. Vande Hey, J. Agnew, J. Richards, R. Dhillon, M. Brettle, "Comparison of mixing layer height retrieved by ceilometer and Doppler lidar," Proceedings of 21st Symposium on Boundary Layers and Turbulence, AMS, June 2014.

J. Vande Hey, J. Coupland, J. Richards, and A. Sandford, "Design and implementation of a divided-lens lidar ceilometer prototype for manufacture," Geoscience and Remote Sensing Symposium (IGARSS), 2012 IEEE International, pp. 5002–5005, 22–27 July 2012.

J. Vande Hey, J. Coupland, M. Foo, J. Richards, and A. Sandford, "Determination of overlap in lidar systems," Appl. Opt. 50, 5791–5797, 2011.

Field Campaigns

InterACT: Intercomparison of Aerosol and Cloud Tracking, Potenza, Italy, 2013.

HygrA-CD: From Hygroscopic Aerosols to Cloud Droplets, Athens, Greece, 2014.

Professional Societies

IEEE (2007-present).
Geoscience and Remote Sensing Society.
Women in Engineering Society.
Signal Processing Society.

Optical Society of America (2007–2008, 2009–present).

Royal Meteorological Society (2013–present).

Contact Details

dr.joshua.vandehey@ieee.org
http://www2.le.ac.uk/departments/physics/research/eos/joshua-vande-hey.

MIX
Papier aus verantwortungsvollen Quellen
Paper from responsible sources
FSC® C105338

If you have any concerns about our products,
you can contact us on
ProductSafety@springernature.com

In case Publisher is established outside the EU,
the EU authorized representative is:
Springer Nature Customer Service Center GmbH
Europaplatz 3, 69115 Heidelberg, Germany

Printed by Libri Plureos GmbH
in Hamburg, Germany